The Perfume Kit

CREATE YOUR OWN UNIQUE FRAGRANCES

The Perfume Kit

CREATE YOUR OWN UNIQUE FRAGRANCES

CHARLA DEVEREUX

and BERNIE HEPHRUN

HEADLINE

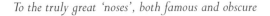

To the truly great 'noses', both famous and obscure

First published in 1995
by HEADLINE BOOK PUBLISHING

1 3 5 7 9 8 6 4 2

British Library Cataloguing in Publication Data

Devereux Charla
Perfume Kit: Create Your Own Unique Fragrances
I. Title II. Hephrun, Bernie
668.54

ISBN 0-7472-1527-8

AN EDDISON·SADD EDITION
Edited, designed and produced by
Eddison Sadd Editions Limited
St Chad's Court
146B King's Cross Road
London WC1X 9DH

Phototypeset in Perpetua using QuarkXPress on AppleMacintosh.
Origination by Create Publishing Service Ltd, Bath.
Book printed in Great Britain by BPC Paulton Books.
Oils supplied by Butterbur & Sage Ltd.

HEADLINE BOOK PUBLISHING
A division of Hodder Headline PLC
338 Euston Road
London NW1 3BH

CONTENTS

INTRODUCTION

Your *Perfume Kit* consists of a set of seven perfume complexes; Musk, White Flowers, Woody, Green Chypre, Modern Floral, Modern Fruits and Rose Accord, and a manual which tells you how to use them. They have been especially selected to allow you to create many beautiful perfumes across the wide spectrum of modern scent. Although you will be anxious to begin immediately, before you start, take the time to read about the history of perfume and the principles which govern its blending. The manual offers a number of formulas to follow as well as outlining the basics. Chapters One and Two talk briefly about the sense of smell and the history of perfume; Chapter Three provides a valuable look behind the scenes at how perfumes are blended, while Chapter Four details how fragrances are classified. Chapter Five offers many exciting ways to use your *Perfume Kit* to make your own wonderful fragrances while Chapter Six tells you how to extend your repertoire.

A UNIVERSAL OBSESSION

Perfume has always been mysterious and special. Not only does a beautiful fragrance have the power to impart charm and distinction to its wearer, it also exerts power over the mind by evoking memories, stimulating the imagination and influencing emotional and spiritual states of consciousness. The universal obsession with perfume is indisputable; it is no longer regarded as an occasional luxury, for its magic is is now an everyday indulgence. Almost everyone uses perfume in one way or another, and whether comforting, exciting or relaxing, many feel naked without it. When you wear a particular perfume, you are saying something about who you are. Although using perfume can be a part of the daily routine, it can also be a more conscious procedure. By learning more about the nature of perfume and, in particular, the enormous range of fragrances available in the perfume family, you will discover how to choose, use and create the perfumes that are most suited to you.

Fragrance is not something invented, but has existed since the beginning of time. Perfume has sometimes been referred to as the soul of the flower, and using this analogy some philosophers have suggested that the soul of man is in the olfactory nerve, because smell is considered the most ethereal of the senses. In Chapter One we discuss the sense of smell and look at the various sources of fragrance in nature, delving briefly into the psychology of smell.

Sweet smells have often been associated with things sacred and pure, while evil is suggested by foul odours. The Elysian Fields, the Greek idea of heaven, had gates of cinnamon and a river deep enough to swim in which gave forth an odorous mist that encompassed the whole area. Perfume was looked on as a divine attribute in early religions and the secrets of perfume-making were guarded by the priests. Egyptian priests, for example, practised the manufacture of perfumes as one of their religious mysteries, using fragrances in the worship of their gods as well as in the funeral ceremonies of their great kings. Over time, more people made use of perfume as knowledge concerning its preparation became more widely known. In Chapter Two the origins of perfume are considered and its fascinating history revealed – from its early use in religious ceremonies to its present-day status as a multi-billion dollar industry where scientists are now able to capture the fragrance of a flower and reproduce it synthetically.

The art of perfumery relies on nature as its model which gives the perfumer an almost unlimited number of possibilities. The first known person to create an essential oil from a flower by means of distillation was an alchemist. Early alchemy sought the transmutation of all matter, not just the transformation of base metals into gold. Some alchemists were in search of a universal elixir which they dreamed would restore youth and prolong life. Perhaps the distillation of oils from flowers might well have been an attempt to capture the romantic essence that exudes from the plant and so captivate those who smell it. Indeed, today's perfumer needs to have a working knowledge of chemistry, be artistic, poetic and understand psychology. Quite a tall order!

Chapter Three describes the work of the perfumer and the materials used in blending. The various processes involved in creating a new scent are discussed, including the concept of notes, which derives from a perceived connection between perfume and music.

As you will see from the charts on pages 104–10, there are a vast number of perfume fragrances, and this number increases each year. While most of the perfumes mentioned in these charts are available today, a few have been discontinued, but we include them because of their historic value. Chapter Four discusses the classification of fragrances and the various systems that have been devised to make sense of commercial perfumes. The different characteristics the categories are considered to present, the psychology behind these classifications and how they reflect the image we wish to project are all explored in this chapter.

In Chapter Five we take a close look at each of the perfume complexes in your *Perfume Kit*, providing ingredients, information and background to each of the complexes. A wide range of formulas is suggested for your experimentation, and after you have become familiar with the techniques of blending, you can make up your own formulas. There are many ways in which the perfumes that you create can be used and a host of ideas is provided, offering instructions for making a range of cosmetics including shampoo, skin cream, bubble bath and lotions.

Chapter Six provides a further range of perfume formulas for classical toilet waters and Eau de Colognes as well as giving a number of formulas for feminine and masculine perfumes. We also look at how to create perfumes by using essential oils, which will allow you to expand your repertoire of fragrant choices and help you develop a range of original scents that truly reflects the real you!

Chapter One

THE BASIC SENSE

*Tell me what you love to smell and I'll tell you
what you are.*
DR IVAN BLOCK

All life is a journey through a maze of odours, each one affecting us in different and varying degrees, according to psychologist Dr Ivan Block. Our language is rich with words that describe what we can see, what we can hear and what we can feel through touch. But when looking for words to describe a fragrance, we are forced to use analogies even though smell is the most direct of all our senses.

Of the five senses, smell is the least valued and understood in Western culture in spite of the fact that our ancestors had a much more highly developed sense of smell than we have today. For most animals it has always been and remains the most important of the senses, crucial for hunting, marking and recognizing territory, for finding a mate and recognizing offspring. For humans, the sense of smell has a profound effect on our ability to taste and appreciate food, it provokes memory in a way no other sense can, and it is essential to the bonding process between mother and baby.

OUR UNDERVALUED ASSET

The average person has the ability to register as many as 10,000 different fragrances and can classify single odours into one of at least sixty categories. Considering that only four types of taste can be recognized helps put this into perspective. It means that the other flavours we think we are tasting are actually odours!

Everyone from time to time experiences what we call a loss of taste as a result of having a cold, but what we are actually experiencing is a loss of smell which renders most of what we eat tasteless. For the majority of people this is temporary; however, for people who suffer from *anosmia* (complete loss of the sense of smell) it is an everyday occurrence. Anosmia, from Latin and Greek meaning without smell, can be caused by an allergy, infection or accident resulting in partial or complete loss of smell. Some people are born with this condition. If the olfactory nerve is severed as a result of a concussion or other accident, it results in permanent damage. For those who suffer from anosmia, certain situations can become life-threatening: not being able to smell smoke or putrefied food can be very serious indeed.

HOW IT WORKS

The mucus membrane in the nose contains about ten million specialized cells which respond to different scent chemicals. What we experience as a smell depends on the combination of cells that is stimulated by the chemicals carried on the air which we breathe. The receptor organ for smell is the olfactory lobe, which is situated on the undersurface of the brain's frontal lobe. Part of the olfactory lobe is the olfactory bulb, an oval mass located at the beginning of the olfactory tract which splits in two, with both roots being continuous with the two extremities of the limbic lobe. The undersurface of the olfactory bulb receives the olfactory nerves (or first cranial nerves) of which there are about twenty.

Upon detection of a smell, the olfactory bulb signals the cerebral cortex and sends a message

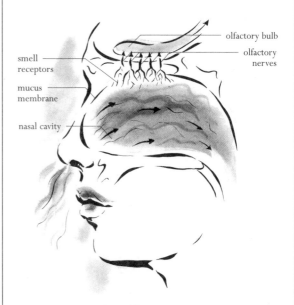

smell receptors

olfactory bulb

olfactory nerves

mucus membrane

nasal cavity

The Olfactory System

to the limbic system which is the emotional centre of the brain. The limbic system in turn is connected to the hypothalamus which has control of the body's hormonal system in addition to being associated with the establishment of memory patterns, mood, learning and emotional states. The olfactory sense is unique among the senses in its connection to the limbic system; because of this it is believed that the effects of odour on mood and emotion may not be consciously recognized. While it takes a mere eight molecules of a substance to trigger an impulse in a nerve ending, forty nerve endings must be triggered before we consciously smell anything.

Smell is dependent upon the release of volatile microscopic particles into the atmosphere where evaporation can take place. This means that if evaporation does not occur then a scent will not be able to be detected.

PHEROMONES

When perfumes were only produced in limited amounts, some animalic essences were used in their production – musk, ambergris and civet are the most well-known. Some of these fragrances come from the anal sacs of the animals. They were used in highly diluted form and today in perfumery have been replaced with synthetically reproduced essences. Most of these smells are rather repulsive, yet when diluted they can be truly exquisite. What is it about these scents that is so appealing to humans? It seems that there is a connection between the geometric shapes of molecules and the odour sensations which they produce. These animal scents assume the same

chemical shape as a steroid and when smelled the response is the same as the response to human pheromones. The word 'pheromone' is from the Greek *pherein*, 'to carry', and *horman*, 'to excite'.

Pheromones are a somewhat recent discovery made by researchers studying the behaviour patterns of ants. It was found that ants leave odorous trails for others in their colony to follow which could lead, for example, to a food source. It was subsequently found that most insects released chemicals to attract mating partners. Unlike hormones, which are regulatory substances produced in an organism and transported in tissue fluids such as blood to stimulate cells or tissues into action, pheromones are chemical substances which are secreted and released by animals for detection and response by other animals, usually of the same species.

There are two ways in which pheromones work in insects. They can produce an immediate and reversible change in the behaviour pattern known as a 'releaser' effect. This means that the chemical substance seems to produce a single specific response which is mediated by the recipient's central nervous system. This serves a number of functions, with sex attractants high on the list. If, on the other hand, the principal function of the pheromone is to trigger a chain of physiological events in the recipient, it has what is called a 'primer' effect. These physiological changes will give the organism a new behavioural pattern which will be activiated by appropriate stimuli. Pheromones producing a releaser effect are very common among animals.

Evidence has shown that humans can, at least,

be physically influenced by pheromones from other animals. For example, smelling musk can trigger a hormonal change in women producing shorter menstrual cycles. In recent years the concept of human pheromones has been under close scrutiny. Much research has been done on the subject, yet there is still a great deal to investigate. The smell of other people is always a matter of interest and yet it is a subject which appears to have missed the attention of the serious researcher. As a result the study of human pheromones is still in its infancy.

SMELL AND MEMORY

The effect of smell is immediate because it can trigger emotions before we even have time to think about them. A familiar smell can be linked instantaneously to a past event even though the present circumstance in which the smell is produced is totally unrelated. Smell is directly linked to memory. A certain smell may be unpleasant to someone, not necessarily because of the odour, but because of the memory it provokes. This can be true for a particular perfume as well. If you have bad memories of a particular teacher who happened to wear one perfume all the time, chances are that when you smell the same fragrance again, even many years later, you will wrinkle your nose at the smell. This response to the odour can occur even before you remember the person in question!

But this connection with smell and emotion can also work the other way around. You may remember being told as a child that the worst thing you can do if confronted by an unfamiliar animal is to show it that you are afraid. What this really means is that your sudden change in emotion will trigger a change in your body odour which will literally be sensed by the animal.

HUMAN SCENT

In humans there are two types of sweat glands. The first are the exocrine glands which are found all over the skin and play an important role in excreting sweat as part of the heat regulatory process. The second type of sweat glands are the apocrine glands which occur in the hairy parts of the body, particularly the armpits and groin, but are also found on the face and chest area. These glands develop in the hair follicles and appear after puberty. Although the function of these glands is not fully understood, their secretions seem to be dependent on endocrine activity which in itself is sensitive to emotional or stress situations. It is believed by some that the smelling of each other's faces is what makes kissing so pleasant. Indeed in some cultures including Siberia, India and Borneo, the word for 'kiss' actually means 'smell'. Smelling the face or hands seems to be a universally practised form of greeting. This may be one of the reasons why the perfuming of gloves became so popular in Europe several centuries ago (*see page 31*).

The secretions from the apocrine glands are naturally odourless. The strong odours associated with sweating result from the action of bacteria on these secretions. The hair in these areas helps to disperse the scent by providing a nutritive jungle for the bacteria living in the skin; it also increases the area available for the evapora-

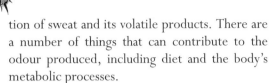

tion of sweat and its volatile products. There are a number of things that can contribute to the odour produced, including diet and the body's metabolic processes.

Unlike the rest of the animal kingdom which consciously makes use of natural scents in a number of ways, we humans use manufactured fragrances to mask our body odours.

It was Theophrastus, the Greek 'father of botany' who recognized that everything – plant, animal or inanimate object – has an odour peculiar to itself, something that most animals are aware of. It is this scent that allows a bloodhound to 'follow the trail' once it has smelled an item of clothing that has been worn by the person being pursued. Each day we individually shed about 50 million skin cells. These rise as a result of the warm air which is produced by the body, and then fall to the ground forming a microscopic trail. This is what the bloodhound follows.

Certainly smell was of prime importance to our ancestors living in the Palaeolithic period – the Old Stone Age – for survival. The hunter required a keen sense of smell to help track his prey and also to help keep him and his family safe from danger. No doubt a keen sense of smell also helped the women gather the roots and plants which were often the staple of their diet. These people led nomadic lives, moving around from season to season in order to have adequate supplies of water, game and plant foods. Body odour may well have been another of our Palaeolithic ancestor's assets; human smells may have acted as boundary markers thus keeping away dangerous or predatory animals.

At the beginning of the Neolithic period, our ancestors stopped their nomadic lifestyle, established settlements and began to cultivate crops and keep animals. This marked the point at which the reliance on the human sense of smell began its decline.

CULTURAL DIFFERENCES

Although there has been very little research done comparing human body odour from culture to culture, there is an abundance of anecdotal evidence which refers to this. Of course it makes common sense that whatever is ingested will have an effect and different cultures have different dietary preferences which can contribute towards a particular smell being associated with a particular race of people. The ability to perceive a particular smell can disappear as a result of sensory adaptation; the individual's sense of smell becomes less and less able to perceive a particular fragrance the longer he or she is exposed to it, and so these types of ethnic smells are only really apparent to someone from a different ethnic background who is encountering the smell for the first time.

The abundance of food choice in modern society does tend to alter some of the more traditional dietary patterns, with, for example non-meat eaters having a different smell than those who eat meat at most meals. It may be of interest to note here that there was a time in Japanese history when a man who had a strong body odour would be disqualified from military service since it was very rare for Japanese people to have much of a body odour. Perhaps diet had something to do with this; until 1856, when beef was first slaughtered in Japan, meat was not part of the Japanese diet.

BONDING AND BODY SCENT

There is an odour-bonding that occurs between mothers and babies during the first few hours after birth which enables the mother to recognize her young. This bond is so strong in animals that a ewe, if she has lost her own lamb, will actually accept another lamb as her own if the dead lamb's skin is tied on to it. If, as happens with some emergency births in humans, the mother is not given her baby for several hours after the birth, it is possible that this bond may not develop in the same way. However, a baby can also acquire associations with artificial odours as well as natural ones: for example, the smell of its mother's perfume, providing she regularly uses one fragrance. It could be that using a variety of scents might be confusing for a new baby, and may cause the child to suffer from a desensitized sense of smell.

It takes the human baby about ten days before it can identify the smell of its mother's breasts, and the close contact between mother and baby can make their smells similar. In experiments done with T-shirts, people who are unrelated can match a T-shirt belonging to a baby with the one belonging to its mother. These 'family odours' are common amongst animals. Animals that travel in packs will merge their individual smells, thereby producing an identifiable group odour which allows them to easily detect the scent of an intruder.

SEXUAL RESPONSES TO ODOUR

Vaginal odours have a strong effect on many animals. A male dog can easily tell when a female dog is on heat by the smell she emits from her vagina as well as her urine. Indeed, for mammals, the smell of the female vagina is a powerful aphrodisiac. Likewise, the male animal odour can affect the sexual cycles of their female counterparts. For some reason grizzly bears are attracted to the vaginal scents of menstruating women; something to remember when camping out in the wilderness! But because most humans today use an array of bottled fragrances, and the smell of natural body odour is considered to be unpleasant as well as primitive, it is difficult to determine what natural instincts we may have that could be influenced by natural body odour.

It is surprising to many people to learn that it was only in the late nineteenth century that natural human body odour actually started to be considered unpleasant and antisocial. This was a marked change to earlier times. For example, in Elizabethan times a woman would give her lover a 'love apple' to inhale, which was a peeled apple that she had kept in her armpit until it was saturated with her sweat. Today, however, it is rare to find anyone who does not make use of some kind of fragrance to mask body odour, even if it is just soap.

It seems that our brains do not seem to register human pheromones as smells, even though the human nose seems to be sensitive to odours of sexual significance. These smells often have a pronounced physiological effect without the individual consciously perceiving it.

It has been noted that when women live in close quarters over a period of time, their menstrual cycles become synchronized. The latest research indicates that this is caused by a pheromone in sweat. Whatever the cause, it happens on an unconscious level. This means that in-depth types of communication can be going on without us being aware of it. This could eventually prove to be a rather large qualification to what we blithely term 'free will'.

OTHER HUMAN ODOURS

Severe metabolic disturbances can cause a high level of ketone bodies in the blood. These ketone bodies are produced during the metabolism of fats, and can cause a sickly sweet odour in breath or urine. This is a condition found in a particular form of diabetes and can also occur as a result of fasting. Schizophrenia is often accompanied by a distinctive odour as well. The point is that the body will give off certain smells which could be used as warning signs that something is wrong. Surely this is a skill that could be further developed for diagnostic purposes?

Pheromones on the other hand can be thought of more as natural aphrodisiacs, ones that we no longer consciously recognize. Aphrodisiacs (named after Aphrodite, the Greek goddess of love and beauty) are substances that arouse and heighten sexual desire. Although sometimes substances are ingested with the hope that they will have an effect on the libido, there is little doubt that the use of fragrant substances evokes feelings and stimulates the senses, providing an aura of sensual imagination.

It is also of interest that a significant number of people who suffer from smell disorders also suffer from a reduced sex drive. For women, in particular, the smell of a lover is perceived to be a sexual stimulant.

MOTHER NATURE'S SCENT

The natural world is rich in both pleasant and unpleasant odours, yet even unpleasant odours may find a place in modern perfumery when used in minute amounts. The plant kingdom reveals a fantastic number of pleasant smells, as a walk around a garden or along a country hedgerow will testify. These 'floral notes' can be numbered in the thousands. But flowers alone are not entirely responsible for the wonderful array of scents. If you crush a few leaves of eucalyptus or cypress, they both reveal rich and distinctive fragrances. Eucalyptus has a familiar nose-clearing effect, and cypress has a rich, 'green' scent which may suggest a wooded area. A walk through a forest can have a profoundly calming effect on the emotions as the odours take effect. The fresh

Eucalyptus

Cypress

smell of 'newly mown hay', as you walk through a corn field, illustrates yet another pleasant odour.

However, a flower does not possess a fragrance purely for the benefit of human pleasure. Although a fragrance such as lavender may be delightful to us, its main purpose is to be a defence mechanism for itself and, by luck or planning, for other plants around it. It seems that deer do not like the smell of lavender, which makes it an ideal plant to use as a border around plants that deer might find more appealing. More commonly, plants give off a fragrance to attract insects and birds and encourage pollination. So the same heady fragrance that we are attracted to can be either an attractant or repellent to insects and many animals.

Some plants have developed ingenious ways of protecting themselves. For example, strychnine, a bitter and highly poisonous substance, can be found just under the bark of some plants, which discourages animals from eating it. Other plants produce flowers that lure insects by their smell and then dust the insect with pollen to repel it, thereby encouraging pollination. There are even some carnivorous plants which trap and digest insects or other animal substances.

Flowers, leaves, buds, roots, stems, fruits, seeds, ferns, mosses and trees not only give forth a myriad of odours, but many plants have the capacity to store the aromatic substances which are generated by their metabolic processes. These products are stored in flowers, fruits, leaves, bark and roots, depending on the particular plant. Some plants like marjoram, rosemary and sage have single or multi-celled protuber-ances on the surface of their epidermis where these oils are stored, while others have inner cellular spaces in their tissues where essences may be manufactured and stored. When a disturbance occurs, canals or ducts are formed which release this stored material.

AN INFINITE PALETTE

When one considers the hundreds of thousands of plant species and the many thousands of varieties within a species (such as the rose, for example), it is possible to imagine the vast array of possibilities these materials offer the perfumer's creative palette. In addition to this there are always new flowers and plants awaiting discovery. Orchids, for instance, were once thought to be just beautiful flowers, but there are now many species which have unusual and quite distinct odours that are being examined.

Many plants are categorized as herbs, while others are grouped under spices. Herbs are non-woody seed-bearing plants which die down to the ground after flowering and are used in medicines and as food flavourings. Spices are aromatic or pungent plant substances mainly used to flavour food. There are also many types of resin which are exuded from the stems, bark and leaves of various trees and plants. These often sweet-smelling products have a firm place in the history of fragrant materials. Many resins found their way into Egyptian perfumes and embalming mixtures. Some, particularly frankincense and myrrh, are referred to many times over in biblical history, and were used as valuable incenses, perfumes and medicines.

SOCIAL EFFECTS OF FRAGRANCE

Over the past ten years much research has been carried out on the social effects of fragrance. Two researchers who have contributed to this work are Michael Kirk-Smith and David Booth. In their paper 'Chemoreception in Human Behaviour: Experimental Analysis of the Social Effects of Fragrances', they suggest that 'human responses to odour are acquired and that the conditions for eliciting a response to a particular odour depend, therefore, on the complex settings of previous social experience with odours'. This means that any study involving the effect of odours on human behaviour needs to take into consideration all the social complexities which surround the individual.

Children do not necessarily have fragrance preferences – but they can learn from associations at birth as well as developing traits from their environment based on the fragrances they are exposed to and the opinions expressed by their parents. So a smell that is vigorously disliked by a parent may well elicit the same response in a child, which the child then carries into adulthood. A certain emotional response, depending on the situation, can occur when the particular smell is present. The smell could be anything from faecal matter to the most expensive rose perfume.

Early life experience is another circumstance in which smell preferences may be determined. When a negative relationship occurs between two people, regardless of whether it is within the context of the family or outside the home, it can leave marks that are carried throughout life. So, for example, if someone you meet in a new job situation happens to smell like a teacher with whom you had a very unpleasant relationship, your unconscious reaction to your new workmate may be directly influenced by the bad feelings you had for your teacher! And the reverse is also true: you can have an effect (negative or positive) on people that has nothing directly to do with you! All of these feelings, negative and positive, will be elicited from memories awoken at an unconscious level.

Another trait that is learned is how we think something or someone should smell. In nature there are variations, but we tend to live most of our lives indoors, in most cases between a handful of very controlled environments – home, workplace and car – that are not only temperature-controlled, but to some extent fragrance-controlled as well. We get so used to the overall environment that we rarely smell it at all, unless of course there is a change for some reason. This expectancy in smell contributes somewhat to the success of modern perfumes which never vary in smell because their fragrances can be controlled to the point where they are always the same, batch after batch. When a variation in scent does occur we generally tend to think that, in the case of a perfume, the fragrance has 'gone off'. In other words, if it is not what we expect, then something must be wrong.

Chapter Two

THE STORY OF PERFUME

Incense perfumes, bad smells, and candles illumine men's hearts.
CONFUCIUS

Long before it became common-place to perfume the body, scent played an important role in the religious practices of ancient cultures. The word 'perfume' is derived from the Latin *per fumum*, which means 'from smoke'. Perfume or incense was used as a form of offering and to drive away evil spirits in many ancient societies. Incense was referred to by the Egyptians as 'fragrance of the gods'. In many cultures the fragrance of burning incense was considered the manifestation of the god or goddess for whom the incense was burned.

A number of suggestions have been put forward concerning the origins and the use of incense. Incense might have been used to neutralize, or at least to hide, offensive smells produced as a result of animal sacrifice or when the dead were cremated. Incense was also considered a medium for prayer; as the smoke rose it carried a message to the deity being worshipped. Regardless of its origin, the evidence is that incense was used in all cultures where records exist.

CHINA

From artefacts handed down within the Chinese culture, it seems apparent that the use, and possibly even the distillation, of aromatics were well-known prior to records being kept elsewhere. Incense, often in the form of joss sticks, was burned in temples in the morning and evening. Sometimes special censers were used, or the incense would be placed at the feet of the deity. In the Hall of Ceremonies in Peking incense was burned in twelve large urns in memory of the deceased emperors. In ancient times perfumes were a luxury because of their high cost. Not only did the raw materials have to be imported in many cases, but also the skill involved in creating the fragrances was considered to be a specialized craft in most cultures, and as a result of these two factors perfumes were very expensive. Often the formulas would be kept secret, in many cases by the priesthood who would pass on these closely guarded secrets very carefully from one generation to the next.

Incense burner

Incense made from aromatic woods was used by Chinese fishermen to propitiate the deities prior to leaving on their boats. It was commonplace to have incense burners in the homes, often at shrines to venerated ancestors. Musk was a favourite not only as a perfume but also for its medicinal qualities and was used to treat a number of different diseases.

INDIA

In India, which enjoys an abundant variety of flowers, perfumes have been mentioned from earliest records for both sacred and personal use. According to the Vedas, a fire of fragrant wood was lit at each of the four cardinal points as a sacrifical offering. A herb (believed to have been ginger grass) was used. Sacrifices were offered both as a general mode of worship and to gain the favour of a particular deity.

A particular flower or scent is associated with each of the five heavens in Hindu mythology. The rare blue campac flower, for instance, was associated mythologically with Mount Meru (in reality, Mount Kailash in the Himalayas). Camalata, a rosy flower, was said to possess the power to grant wishes in Hindu tradition, and offering perfumed water to wash the body of the god is an essential part of Hindu worship.

Sandalwood is a favoured fragrance which has been used principally for religious purposes from time immemorial in India. It was mentioned in the Nirukta, an old Vedic commentary, which was written no later than the fifth century BC. Sandalwood was used for embalming the bodies of the princes in the ninth century in Sri Lanka, and was also made into incense or joss sticks for use in the temples. Sandalwood, along with other fragrant ingredients, was burned during Hindu marriage ceremonies and during cremations. Spikenard was also highly prized. Other perfumes from India included patchouli, cassia and cinnamon. Urgujja was a frequently used perfumed unguent (ointment) which consisted of attar of rose, essences of jasmine, sandalwood and aloes wood.

Ghazepore, situated on the north bank of the sacred River Ganges, was the key centre for the manufacture of essence distillation in India. The process involved placing the petals being used into clay stills with twice their weight of water, then exposing them to the fresh air overnight. The otto (essential oil) would be skimmed off the surface where it had congealed. Sandalwood shavings were sometimes added to the petals which helped to facilitate the extraction of the otto, but this would contaminate the natural scent of the flowers.

Another means of extracting essential oil was to place gingle oil seeds (sesame seeds) in alternating layers with the fresh flowers in a covered vessel. The layers would be renewed a number of times prior to the seeds being pressed. Evidently this oil absorbed the scent of the flowers better than any other.

ARABIA

From very early times, Arabia was the chief source of such fragrant plants as olibanum (frankincense), myrrh, jasmine and rose. A  tenth-century Arabian doctor called Avicenna (Ibn Sena) is an important figure in the history of perfume, for not only was he the first to study and apply the

Myrrh, jasmine and rose principles of chemistry to perfume-making, but he is also credited with inventing the art of extracting essential oils from plant materials by means of distillation. The rose (*Rosa centifolia*) was the first flower whose essence was captured in this way and as a result rosewater soon became prized throughout Europe and the East. It had many uses in perfumery, in cooking and in medicine but it was also used in censers (sometimes in combination with other fragrant materials) to create a smoke that was wafted over visitors as a gentle reminder to them that it was time to make their departure!

Avicenna's most important work was the *Q'anun* which encompassed the work of many of his predecessors including Galen and Rhazes. Indeed, it was the investigative work of such scholars as Avicenna in the fields of botany, chemistry and alchemy which laid the foundations for later scientific research.

EGYPT

We know that perfumes were used extensively in Egyptian culture from the evidence of various tools found in tombs and from the written records in hieroglyphics which have

This Egyptian carving illustrates how lilies were gathered and processed to extract their essential oils.

been deciphered. An early carving which is now in the Louvre shows women gathering lilies and placing them in a device which extracted oil from them. Fragrance was held in such high regard that there was even a god of fragrance, Nefetem, who was shown rising out of a lotus in some records. This flower appears in many scenes in stone hieroglyphics on tombs and temples as well as in papyrus records. There was an unbroken tradition of priests recording and handing on the knowledge of plants. Most temples, such as the temple of Hathor at Denderah, had special rooms or laboratories where oils and unguents were prepared. This famous temple

was a healing centre for many centuries.

Perfumes were used in three principal ways in ancient Egypt: for offerings in religious ceremonies, for embalming and in personal toiletry. A school of herbalists was established at Heliopolis in about 2800 BC where plants were studied. It would not have escaped their investigation that plants could be used in a variety of useful ways. Temples soon incorporated a 'laboratory' within the sacred precinct where studies and blending were carried out. One spectacular record of this is the preparation of the sacred perfume known as *Kyphi* which covers all of the walls of the laboratory in the temple of Horus at Edfu. As an offering, incense symbolically represented the interaction between the human and

KYPHI

Plutarch described Kyphi as a mixture containing sixteen herbs and resin:

Acorus calamus (sweet flag), *Andropogon schoenanthus* (aromatic rush), *Pistacia vera* (pistachio resin), *Cinnamomum zeylancium* (cinnamon), *Mentha sativa* (mint), *Cyperus esculentus* (cyperus grass), *Commiphora myrrha* (myrrh), *Lawsonia inermis* (henna), *Elettaria cardamomum* (cardamom), *Juniperus communis* (juniper berries), *Crocus sativus* (saffron), *Pistacia terebinthus* (turpentine resin), *Cytisus lanigerus* (aspalatus), *Iuncus odoratus* (rush), *Cinnamomum cassia* (cassia), *Boswellia thurifera* (frankincense), *honey and raisins steeped in wine.*

divine spheres and so it was magically significant in transforming the deceased into a divine state. Incense was used for daily ceremonies as well as major festivals. For example, Ra, the Sun God, ,was honoured three times a day by the burning of incense: resin was used at sunrise, myrrh at midday and Kyphi at sunset. Kyphi was also burned in the home during the night. There seems to have been more than one recipe and, in addition to the record at Edfu, inscriptions have been found on the temple walls at Philae.

A wall painting in a tomb at Luxor depicts a perfumer's workshop, showing the different stages of perfume preparation. Assistants are seen grinding the aromatic materials: one mixes the material with oil in a large pot and another stirs the liquid in a basin that stands on top of a stove while another figure holds a strainer through which a liquid is being poured, most likely the essential oil.

EMBALMING

The use of essential oils, herbs, spices and resins was crucial to the practice of embalming, a cornerstone of Egyptian religious belief. Egyptians believed that the body needed to be preserved in order to ensure the survival of the *Ka* (the spiritual double) the *Ba* (the non-physical aspects of a human) and the *Akh* (the state in which the deceased exists in the afterlife). Such oils as juniper, cassia, cinnamon, cedarwood and myrrh were among the ingredients used in various aspects of mummification. Jars of mixed gums and flowers have been found in tombs, along with censers and lamps which were made for burning perfumed oils. Some of the jars found bear Chinese inscriptions which supports the notion of trade in ancient times. The burning of incense was believed to provide food for the *Ba* and *Ka* of mummies. Many temple inscriptions show the pharaoh offering incense to the gods.

A vase was discovered at a tomb in Luxor which contained some of the original perfumed unguent that was placed there when the tomb was sealed. Chemical analysis revealed that the substance was of animal origin and showed no traces of vegetable fibre, coconut or palm-kernel oils. The ingredients consisted of about ninety per cent animal fat and ten per cent resin or balsam. It was also felt that the preservation of the substance was due to the olibanum or some other gum-resins which are now known to have antiseptic properties and were used by the Egyptians of that period.

ROOM SCENTING

The use of scent and fragrancing was not restricted to embalming practices and ritual – wall carvings allude to the use of perfumes for much more temporal purposes. For example, scenes depicting banquets invariably showed fragrant unguent cones being used, including the blue lotus, a symbol of rebirth, which was reported to have the most exquisite odour. Papyri have been found which describe various herbs and resins such as myrrh, cinnamon and galbanum and some contain recipes for their use.

Indications that trade took place with other cultures to obtain the raw materials for perfume-making exist in various wall carvings. The female pharaoh Hatshepsut (*c.* 1500 BC) who built the beautiful temple of Der-el-Bahari in Upper Egypt, had frankincense and myrrh trees planted in a huge garden in front of the temple. The walls of the temple were decorated with bas-reliefs illustrating the expedition which was mounted to Punt to obtain these trees.

PERFUMES AND THEIR DAILY USE

Two other well-known Egyptian queens, Nefertiti and Nefertari, were dedicated users of perfumed materials. Their palaces had beautifying rooms where baths could be taken and unguents, powders and perfumes prepared and used. The use of perfumes and ointments was a very important part of daily Egyptian life. After washing or bathing, the body was often rubbed with fragrant oils and ointments: the dry desert climate made it necessary to moisturize the skin and massage with oils was used to keep skin supple and elastic. Initially, ointments were obtained from the priests as they alone knew the art of preparation.

It was Cleopatra, however, through her devotion to the art of perfumery and cosmetics who made the most extravagant use of aromatics, raising the appreciation of such matters to a high art. It was said that she took daily baths in perfumed oils and, through the use of aromatics, seduced Mark Antony at their first meeting.

PERFUME PRODUCTION

Three methods of producing perfumed oils were employed in Egypt. The first of these methods, known as *enfleurage,* involved steeping the flowers or aromatics in oils or animal fats until the scent from the materials was imparted to the fat. This fat was often moulded into cosmetic cones for perfuming hair wigs. The second method, known as *maceration,* involved chopping up flowers, herbs, spices or resins into hot oils, then once the fragrances has been imparted, the oil was strained and put into amphora jars or alabaster containers. This method was used principally for skin creams and perfumes. The last method, *expression,* involved putting flowers into bags or presses which extracted the aromatic

Egyptian courtiers, shown here in a second-century BC wall painting from Thebes, tied cones of scented animal fat to their wigs.

oils. Wine was often included in the process and the resulting liquid was stored in jars or pots and used subsequently as perfume.

Perfumes were also manufactured in the form of dry powders or as unguents. Ben oil, from the moringa tree indigenous to Egypt, was often used as a base because it was clear and odourless, did not go rancid and blended well. Castor oil was also used as a base oil for a number of perfumes. Care was taken in the preparation and mixing of perfumes; the order in which ingredients were added and the temperatures used for processing were observed to have an effect on the resulting fragrance. According to Lise Manniche in *An Ancient Egyptian Herbal*, one of the most famous Egyptian perfumes was made in the city of Mendes and was known as *The Egyptian*. It con-

sisted of balanos oil expressed from a thorny tree which was once abundant in the Nile Valley, plus myrrh and cinnamon, and was said to last longer than any other perfume. There is speculation that the Egyptians knew about the distillation process, but there is no known representation of distilling apparatus.

It was believed that perfumes deteriorated under excessive heat so unguents were often kept in the coolest of stone containers such as alabaster, but onyx, glass and shells were also used as containers for perfumes. These unguent jars preserved their contents very well; indeed, when Howard Carter opened a few jars some twenty-two centuries later in Tutankhamun's tomb, a fragrant odour reminiscent of the original material remained.

BABYLON

The Babylonians were also great users of perfumed unguents and oils which they stored in glass or alabaster vessels. Oils recorded on their clay and cuneiform tablets include frankincense, cedar, myrrh, calamus, cypress, myrtle and almond. Incense was used during medical treatment to exorcize evil spirits which were believed to cause disease; depictions of this practice were also found on their cuneiform tablets. For a long time Babylon was the principal market for aromatic gums and spices.

Perfumes were held in great esteem and were given as gifts to sovereigns. Inscriptions found on the ruins of the temple of Apollo at Miletus

record gifts of frankincense and myrrh given by the King of Syria in 246 BC. When the Queen of Sheba visited King Solomon, among the gifts she presented him with were oils and spices. (Solomon's famous temple of Jerusalem was build of stone, sandalwood and Lebanon cedar, which almost totally destroyed the cedarwood forests of the Lebanon.) Although the main reason for her visit was to reach an agreement with Solomon over the spice routes, they apparently got on rather well. A descendant of their union was Nebuchadnezzar, the famous king of Babylon mentioned in the Bible.

Nebuchadnezzar built the highly acclaimed

hanging gardens containing cedar, cypress, mimosa, rose and lily for his beloved wife Amytes. These fragrant gardens were an indication of the importance that scent had for the Babylonians; it is not surprising that perfumery was of major significance in Babylon. Liquid perfumes were contained in bottles of glass or alabaster while ointments were stored in boxes of porcelain or chalcedony, a soft stone similar to alabaster. The Babylonians consumed large amounts of aromatics and according to Herodotus they used perfume over their entire bodies. Fragrant cassolettes were kept burning constantly during banquets.

The Hanging Gardens of Babylon were considered to be one of the Seven Wonders of the Ancient World. Formed in a series of terraces planted with cedars, cypress, roses and lilies, the gardens were irrigated by an hydraulic system.

THE HEBREWS

One of the arts acquired by the Jews as a result of their long captivity in Egypt was perfumery; however only priests were allowed to offer incense in the temple. Severe punishment awaited anyone else attempting to do so.

The oil which had originally been used to anoint Egyptian kings was presumably prepared by Moses on his return from Egypt. Ingredients used for this holy oil were myrrh, sweet cinnamon, sweet calamus, cassia and olive oil. The oil was also used to anoint the high priests, the tabernacle and all the sacred vessels and instruments used in religious rituals. Although Jewish kings were anointed, it is not clear whether this was done with the holy oil. Anointing with oil symbolized a change in status, bringing the person being anointed under special divine guidance, either setting them aside from the profane world or returning them to a normal condition. The oil would be poured on the head and allowed to run down on to the beard and garments. In addition to the holy oil, a formula for a holy perfume consisting of sweet spices, myrrh, onycha and galbanum was given to Moses. Even today, the tradition of anointing monarchs continues in the coronation ceremonies of British kings and queens.

The use of perfume as a tool for seduction was employed long before Cleopatra practised her fragrant art on Mark Antony. Judith, a rich Israelite widow, saved the town of Bethulia from Nebuchadnezzar's army by first captivating the besieging general, Holofernes, and then cutting off his head while he slept. The Old Testament says that she 'anointed herself with precious ointment, and decked herself bravely, to allure the eyes of all men that should see her'.

Not only did the Hebrews use aromatic substances for personal and ritual use, but they also used them as seasoning for meats and wine.

GREECE

The spread of Alexander the Great's empire provoked considerable changes in the main spices and aromatics trade routes as a result of his conquests. The Greeks adapted many of the customs and manners of the Persian Empire conquered by Alexander; he had the floors of his apartments sprinkled with perfumes and myrrh and other aromatics were burned in his halls. The Persians were noted for their perfumes, particularly for the place perfumes held in everyday life (at least for those at court). Persia had long been regarded as the home of rose perfume (Shiraz) and the whole country contains some of the most fragrant plants including frankincense, myrrh, jasmine, and of course, the rose. The specific locations and exact sources of these plants were skilfully concealed in order to keep prices high. The Greek term for perfume, *myron*, may possibly be derived from the word 'myrrh', one of the best known aromatics of the time.

Greek scientists, particularly the Naturalists, studied plants and their effects. Hippocrates, the father of medicine, was not merely a physician — he was also a herbalist who studied and used plants as medicines. Theophrastus is perhaps best noted for his work entitled *An Enquiry into Plants*, written about 295 BC. He also wrote the first known work devoted to perfumery, a short treatise entitled *Concerning Odours*. Much that was written by him remained the most important botanical works throughout antiquity and into the Middle Ages. His writings contain information on the nature of odour in plants, oils as vehicles of perfumes, the properties of perfumes and how to make perfumes using the ingredients which were available in the classical world. According to Theophrastus, plants used for perfume included cassia, cinnamon, cardamom, spikenard, balsam of Mecca, aspalathos, storax, iris narte, kostos, saffron, crocus, myrrh, kypeiron, ginger grass, sweet flag, sweet marjoram, lotus and dill.

MEDICINAL USES

Perfumes were classified as medicines in Greek culture and in fact the terms 'perfumer' and 'apothecary' were used interchangeably. Recipes for some of the more popular essences were found inscribed on marble tablets in the temples of Aesculapius and Venus. Priestesses of the various deities dispensed preparations associated with a particular deity. These preparations were said to be endowed with particular virtues.

A story to illustrate this point refers to a fair young maiden named Milto who would deposit garlands of fresh flowers in the temple of Venus. She developed a tumour on her chin and shortly after had a dream in which the goddess told her to place some of the roses from the garland on the tumour. The young maiden followed the instructions and the tumour soon vanished. Milto eventually became the favourite wife of Cyrus of Persia. The rose, known as the queen of flowers, has been highly prized ever since, enjoying fame not only as a magnificent fragrance, but as a powerful antiseptic as well.

Aeron of Agrigentum is credited with being the first to purify the air with fragrance by throwing aromatics into fires that were started specifically for that purpose. This act curbed the plague in Athens in 429 BC.

PERFUMERS AND THEIR PERFUMES

There were a number of celebrated perfumers in Athens, and the most skilled among them were women. One particular perfumer, Apollonius of Herophila wrote a treatise on perfumery in which he named the places where he felt the best individual aromatics could be found. He suggested, however, that the superiority of a perfume was due to the material and the blender, not the place in which it was grown. A particularly popular perfume of the time was *Megaleion* (also called *Megallium*) which was created by Megallus. Not only was it long-lasting, but it was also valued for its ability to relieve inflammation from wounds. This perfume was made from cassia, cinnamon, myrrh, burnt resin and oil of balanos. Other perfumes of the time included *Metopium*, *Mendesion*, *Nardinum*, *Susinum*, *Rhodinum* and *Royal Unguent*. Names given to perfumes either reflected the name of the inventor or gave an indication of the ingredients from which they were prepared.

Calamus was one of the principal fragrant plants used in Greek perfumes, along with storax. Ancient storax which was native to the Mediterranean region, was a sweet oleo-resin which was also known as Jewish frankincense.

The Greeks made dry perfumes by simply reducing the various ingredients to a powder substance and mixing them together. Compound perfumes were made by bruising and placing different aromatic gums in a container. Periodically the container would be opened and the gum giving off the strongest smell would be removed to ensure that no single aroma dominated. They also discovered that strong fragrances such as myrrh oil became sweeter when mixed with wine. In turn, aromatic perfumes were added to wine to produce sweet-flavoured wines. This practice continues today with the addition of resin to Greek white wine to make retsina.

Both Corinth and Chaeroneia were major perfume production centres, while Alexandria was the big market for gums, aromatics and perfumes. Pliny describes in detail how workers in the perfume factories had to remove all their clothes before leaving work as a security measure because the materials used to produce the perfumes were so costly. Much of this manufacturing took place in the Canopus district of Alexandria, and because perfumes were put into fancy bottles and jars, a flourishing industry in glass and stone containers arose in the same district. By the third century AD taxes were being paid by *aromatopyles* (dealers in perfumes) and *myropyles* (dealers in unguents) at Arsinoe.

RELIGIOUS ROLES

Perfumes were used extensively in funeral ceremonies. The bodies of the dead were usually burned and incense and wine would be thrown on to the fire, usually by the friends of the deceased. The remaining bones and ashes would be collected and mixed with fragrant ointments prior to putting the remains into a funeral urn. Flowers and perfumes were often strewn over the tombs of the deceased. For those who could not afford such lavishness, perfume bottles were painted on to the coffins.

Although the largest consumption of aromatics was during festivals and banquets, in Homeric times it was customary to offer guests a bath and fragrant oils before eating. Aromatics were also present during the meal, and in some cases guests were given garlands of flowers for their hair. One of the virtues attributed to perfumes (rose in particular) was that, when worn, a person could drink as much wine as he or she wanted, without feeling ill afterwards. In some cases the perfume was actually added to the wine to improve the taste!

THE ROMANS

Rather ironically, Greek philosophers, for the most part, condemned the use of perfumes. Still, as Greek culture spread westwards, becoming the basis of Roman civilization, so did the lavish use of perfumes. The Romans merely adapted Greek and Egyptian ideas and expanded upon them. Roman civilization, however, produced a number of very fine authors who studied plants. Cato, Varro, Columella and Galen all mentioned the preparation of perfumes by digestion of the flowers and spices using olive oil or sesame oil.

In his *Historia Naturalis*, Pliny wrote about the use of cosmetics for preserving and improving the complexion. He also provided descriptions of aromatic plants including information on where they grew.

The perfumers of Rome were called *unguentarii* and were primarily of Greek origin. Some ingredients were extracted from flowers grown in Italy but most of the perfume materials were imported, mainly from Egypt and Arabia. There were three kinds of perfumes used by the

Romans: *hedysmata* (solid unguents), *epichismata* (liquid unguents) and *diapasmata* (powdered perfumes). There were many unguents, some named after the ingredients used or place of origin, and others were named after the peculiar circumstances surrounding their production. As new unguents were made they would assume popularity over already-existing ones.

Simple unguents perfumed with one fragrance included *Rhodium* from roses, *Crocinum* from saffron, *Melium* from quince blossom, *Metopium* from bitter almonds, *Narcissinum* from narcissus flowers and *Malabathrum* from cassia. Compounds were also produced. One of the popular compounds, *Susinum*, was made from lilies, ben oil, calamus, honey, cinnamon, saffron and myrrh. Another called *Regal* consisted of twenty-seven ingredients. Other popular perfumes were *Roses of Poestum, Onegalium, Rosemary, Sweet Smelling Rush Cinnamomum,* and *Orris and Balm of Gilead,* which was made from the gum of the plant amyris grown on the mountains of Gilead. These unguents were applied to the body, including the soles of the feet. Sweet odours were also applied to beds, clothes and the walls of rooms.

MEDICINAL AND COSMETIC USES

The herbalist Dioscorides was another important figure who studied and wrote about plants and their uses. A physician in one of Nero's legions, he gathered a good deal of information in the course of his work. In his *De Materia Medica*, he discussed the components of perfumes and their medicinal properties and provided detailed perfume formulas. His writings were the standard work for some fifteen centuries. Among the more unusual items popular with Roman ladies was a special ointment made of sweat scrapings from the bodies of gladiators. The scrapings were mixed with the same oil used to massage the gladiators' bodies, and sold at a high price as an aphrodisiac called *Rhypos!*

THE BANQUET AND THE BATHS

Nero used perfumes lavishly and it is said that he burnt more than a year's supply of spices and perfumes from Arabia on the funeral pyre of his wife Poppaea. There were movable ivory panels on the walls in Nero's dining rooms which concealed silver pipes from which odoriferous essences were sprayed. Another method for perfuming the air at Roman feasts involved coating the wings of doves with fragrance and releasing them in the dining hall. The perfume would be dispersed in a kind of spray as the birds flew about. Aromatics were also used in the fountains at the amphitheatre to refresh the atmosphere.

The baths became an important part of Roman life, at least for the wealthy who, in some cases, anointed themselves three times a day with rare perfumed oils. One of the rooms in the baths, called the *unctuarium* contained large jars of ointments and essences for the use of the bathers. Shaving was introduced into Rome in AD 454 by a Sicilian named Ticinus Menas who set up a shop near the temple of Hercules, along with other barbers from his country. The elite soon adopted this new fashion and so clean-shaven chins and hair redolent with ointments became popular. The use of perfumes to scent the body became commonplace.

An enchantment with perfumery characterized the period from 50 BC to AD 300. Alexandria, though in Egypt, was still the intellectual powerhouse of the Roman world. It was the centre of glass-blowing and produced beautiful glass perfume bottles (which are copied and sold even today in the Khan El Khalili market in Cairo). The Alexandrian chemists were divided into three schools, one of which was the school of Maria the Jewess, which worked on apparatus for distillation and sublimation. This group included a chemist by the name of Cleopatra who is credited with the invention of the *bain-marie* or water bath. This discovery would have been particularly useful for extraction of aromatics in oil by maceration.

An important culinary work, *The Roman Cookery Book of Apicius,* was written in the first century AD. It incorporated many aromatics in circulation at the time of the emperors in its numerous recipes for fragranced foods and wines. However, the extravagant use of perfumes by the Romans eventually caused a law to be passed restricting the use of perfumery on the grounds that overindulgence could cause a shortage of the fragances for temple use. The austere preachings of the Christians also curtailed the extensive use of perfumes.

LATER EUROPE

The Romans introduced some of the practices of toiletry they had learned from the Greeks into England and France when Rome held power over much of Europe. Prior to the Crusades, however, perfumes were only used by the Church or the very wealthy. Knights returning from the Crusades brought back perfumes such as rosewater from the East and it became customary for the wealthy to offer guests rosewater with which to wash their hands after a meal.

Perfumers were known in France by the twelfth century. They were required to serve a four-year apprenticeship and an additional three years before becoming a master perfumer. When perfumes and potions were created, patents were obtained which were then registered in parliament, with patent records going back to 1190.

When Catherine de Medici married Henry II in 1533, she brought her own perfumer, René, with her from Florence. René's shop became a very popular meeting place for the elite of the time, ushering in the general use of perfumes among the wealthy. Catherine de Medici became known as the patron of perfumery in France, sponsoring the development of the industry in that country. The use of perfumes among the aristocracy continued to increase and it is said that Louis XIV took part in the design of his own perfumes.

Perfumes continued to grow in popularity and during the reign of Louis XV it was considered fashionable to use a different fragrance each day;

evidently a large portion of Madame de Pompadour's household budget was spent on perfumes. And so the lavish use of perfumes continued among the aristocracy up until the Revolution, when such luxuries were severely curtailed. But not for long – Napoleon and Josephine soon re-established this extravagance. He had a fondness for perfumes, especially Eau de Cologne. Josephine preferred stronger fragrances, particularly musk. Perhaps Napoleon's fondness for perfume did extend to Josephine's chosen favourites; he is supposed to have sent her a note that said '*Ne te laves pas, je reviens*' which means, 'Don't wash, I'm coming home'. Towards the end of the seventeenth century perfumery in France began to be studied scientifically and flowers began to be cultivated specifically for essential oil production.

ITALY

In Italy the art of perfumery began early in the sixteenth century when Venice became the centre for trade in aromatic gums and woods. Monks, with their knowledge of both botany and chemistry, were the first to create perfumes. A laboratory for the manufacture of perfumes was set up in 1508 in Florence at the Dominican monastery of Santa Maria Novella where elixirs and skin preparations were made for both cosmetic and medicinal purposes.

The perfuming of gloves, invented by the Marquis Frangipani, became fashionable. Gloves were coated on the inside with pomatums (pomades) scented with ambergris, musk or civet, which softened both the gloves and the hands that wore them. One formula for perfuming gloves is given on page 31.

During the seventeenth and eighteenth centuries, the laboratory of the Dominican monks of Santa Maria Novella became well-known for the scents created there.

BRITAIN

Perfumes, which were mainly imported from France and Italy, came into vogue in England during Tudor times although prior to that time some fragrances had been sold by the guilds of Pepperers and Spicers of London which imported spices from Arabia and other locations in the East. These guilds sold spices, drugs and perfumes in their shops and in lieu of taxes gave the king a quantity of pepper each year.

However, it was not until the reign of Queen Elizabeth I that perfumes came into more general use in England, with many of the applications being adopted from the French. A pair of perfumed gloves was said to have given the queen much pleasure; these had apparently been given to her by the Earl of Oxford who brought her various fragrant things from Italy. The scent became known as the *Earl of Oxford's Perfume*.

Pomander

A FORMULA FOR PERFUMING GLOVES

*450 g (1 lb) roses, 450 g (1 lb) orris,
14 g (½ oz) benzoin, 14 g (½ oz) storax,
170 g (6 oz) calamus, 85 ml (3 fl oz) essence of
citron, 14 g (½ oz) coriander, 170 g (6 oz) girofle,
7 g (¼ oz) lavender, 14 g (½ oz) powder of orange,
113 g (4 oz) rosewood, 14 g (½ oz) santal,
demy muscade (½ nutmeg)*

In common with a number of other members of the court, Elizabeth often carried a pomander (a ball of perfumed aromatics first introduced in the fourteenth century), which could be carried in various articles of jewellery and contained a mixture of fragrances. A favourite scent was made up of one grain of civet and two grains of musk added to rosewater. The vessel in which perfumed materials could be stored also came to be called a pomander. Queen Elizabeth I had a cloak and some shoes treated with perfumes.

Other objects designed for carrying aromatics included the cassolette, a small box with a perforated lid which allowed the aromatic contents to be inhaled. It was usually made of ivory, gold or silver. Vinaigrettes were little boxes made of silver with gilded interiors to prevent corrosion which contained a sponge soaked with aromatic vinegar. They were used as a snuff to help clear the senses and ward off unpleasant smells. Hollow beads of necklaces were filled with aromatic gums and rings had compartments which could hold perfumed powder. The tops were pierced with tiny holes to allow the fragrance to come through.

Vinaigrette

THE ROSARY AND OTHER FRAGRANCING OBJECTS

Moulded aromatic pastilles of a similar composition to those mentioned by Dioscorides (*see page 93 for recipe*), were used by Byzantine Christians as prayer beads which, as roses were the main ingredient, became known as a rosary. These beads were exported all over the Christian world from Cyprus.

The rosary is said to have been introduced by Saint Dominic after he was shown a chaplet of roses by the Virgin Mary. It consisted of a string of beads made of roses which were tightly pressed into a round mould and strung together. If real roses were not available then rose leaves were used. The use of a chaplet of beads for recording the number of prayers recited is of Eastern origin and seems to date back to the Egyptian Anchorites. In Renaissance Florence, the nobility counted their prayers on elegant rosaries of silver, gold or pearl beads coated with fragrant gums which were bound together with silver filigree. Often there was a large scented ball known as an *oldano* which was attached to the end of the beads with a gold chain. These beads were also worn in church as a protective measure against any germs that might be lurking among the many people crowded into the churches for mass.

Other paraphernalia associated with perfume including the perfume pan, used to scent rooms that had been closed off for a period of time, and the casting bottle with its perforated top, which was used to sprinkle perfume on the face, head, hands and clothes.

As mentioned earlier, perfumed gloves became very fashionable in England, with strong scents such as musk and civet generally used as bases for the perfumes. Other customs, such as sprinkling rosewater on clothing, were taken from the East. Cedarwood or sandalwood were kept in coffers with linens to deter moths and fragrantly scent the fabrics. It was also customary to perfume a room by either strewing the floor with sweet rushes, or by burning perfumes in a room to fumigate the sheets. Perfumed bellows were also used for this purpose.

It became common for the wealthy to have a still room devoted to the production of perfumed products such as dry powders, oils, pomatums or distilled waters in their houses. It was customary to collect and keep recipes in a book (which lived in the still room) that contained various formulas for perfumes, domestic medicines, cookery and brewing. These recipes were preserved and carefully handed down from generation to generation.

PERFUME FORMULAS

Perfumes used in the sixteenth century were generally in the form of dry powders, mixed sachets or perfumed water. Dry perfumes were sometimes placed in silk bags which could be tucked in with clothing.

There were a number of books dealing with the use of perfume, containing information and formulas for a wide range of applications including perfumes for powders, lamps, perfumes to combat the plague, and room fragrancers. One popular title was *The Secrets of Alexis*, first printed in Italy in 1555 and later translated into a number of languages. One of the formulas it contained was for *Damask Perfume (see box below)*. Popular fragrances of this time included damask rose, lily, violet, civet, musk, rosewater and marjoram.

Lavender was a favoured fragrance not only for the lovely smell but also because it kept moths at bay. The twelfth-century Abbess Hildegard of Bingen on the Rhine is credited by some with the invention of lavender water, which is still a very popular perfume particularly in England. Lavender was one of the fragrances used to prevent infection from the plague and has long been associated with cleanliness.

Certain aromatics were used for specific purposes. Rosemary was used at both weddings and funerals as well as for medicinal purposes. Indeed, the medicinal aspects of fragrance were often taken into account: it was noticed that in laboratories where perfumes were made, most workers did not suffer from respiratory complaints. Fumigation by burning various plant materials was used during times of plague.

DAMASK PERFUME

*5 grains musk, 2 grains civet, 4 grains ambergris,
4 grains fine sugar, 1 grain benzoin, 3 grains storax
3 grains calamus, 2 grains aloeswood*

The ingredients were beaten to a powder, placed in a perfume pan with rosewater ('two fingers high'), then placed over a fire. When the water evaporated, more was added, and so on for several days.

BODY PERFUMES

It was in the eighteenth century that the practice of using perfume on the body came into vogue. Initially this was done through the use of powder-patches. Perfumed handkerchiefs were also carried. The first known professional perfumer in England was Charles Lilly, who owned a shop in the Strand. He was highly praised for his snuffs and perfumes which were described as uplifting or calming depending on the requirements. In 1822, Lilly wrote a book entitled *The British Perfumer* which records descriptions of materials including musk, ambergris, benzoin and rose otto. Lilly offered ideas on how these various materials could be used. Soon other chemists began to open shops where they distilled aromatic waters including lavender, elderflower and rosemary which were popular at that time. Another famous perfumer, William Bayley, opened a shop selling perfumes in 1739 which he named Ye Olde Civet Cat.

Amidst all the popularity, however, there were those who denounced the use of perfumes as vanity and worse. In 1770 a sponsored act of Parliament stated:

That all women, of whatever age, rank, profession, or degree, whether virgins, maids or widows, that shall from and after such Act, impose upon, seduce and betray into matrimony any of His Majesty's subjects by the scents, paints, cosmetic washes, artificial teeth, false hair ... shall incur the penalty of the law in force against witchcraft and like misdemeanors and that the marriage, upon conviction, shall stand null and void.

According to the *Perfumer's Manual* written by Madame Celnart in 1834, there were many perfumers practising at that time, and, in the second half of the nineteenth century perfumers began to be publicly recognized. Alas, this was not to continue. In the world of commercial perfumery which started to expand at the turn of the century, many companies did not want their success to be attributed to an individual except possibly the proprietor of the concern, and as a result most perfumers became anonymous workers.

Throughout history, perfumes were used for their fragrance as well as their healing properties. Along with the increasing popularity of perfumes came the charlatans who claimed to have the special remedy for all types of ailments in the form of powders, elixirs and pills. Eventually laws were passed making this practice illegal. Perhaps it was from this point that perfumes began to be used solely for fragrance.

A Brief History of Modern Perfumery

It was Eau de Cologne, based on alcoholic extracts, which initiated modern perfumery. The advantage was that alcohol-based formulas evaporated quickly, leaving a sharp, astringent perfume which refreshed the skin while leaving it pleasantly scented.

When first formulated by Jean-Paul De Feminis in 1690 in Milan, it was called *L'Eau Admirable* and was made primarily of neroli, lemon and bergamot with a little lavender. It had the scent of a bouquet, rather than being based on a single fragrance. Apparently, De Feminis passed on the formula to his nephew, Jean-Antoine Farina who lived in Cologne. Farina produced the fragrance there with some modifications. A relative of Jean-Antoine Farina, Jean-Marie Farina came to Paris in 1806 where he produced the formula which became known as *Eau de Cologne*, since he came from Cologne, Germany. There is, however, another version of the story. Two brothers, also with the last name of Farina, claim that they immigrated to Cologne from Italy around 1709 and created various waters, one known as *Aqua Admirabilis*. During the Seven Years War French soldiers stationed in Cologne liked the perfume so much that they brought it back to France with them, where it was very well received and given its name, *Eau de Cologne*. Later, in 1806, Jean Maria Farina, a nephew of the two brothers, brought the formula to Paris. Regardless of which story is true, every perfume house of that time put out their own formula under the name of Eau de Cologne. The Farina perfume house was eventually taken over by Roger et Gallet where the formula is still produced under the same name. Paris, already a leader in the world of fashion, was to become the perfume capital, with many of the great couturiers becoming sponsors of perfume fragrances, a practice which continues today. There are many formulas for Eau de Cologne, a number of which are trade secrets (*see page 91*).

OTHER EARLY PERFUMES

The first known alcohol-based perfume, made primarily from rosemary and wine alcohol was called *Hungary Water*, after Queen Elizabeth of Hungary for whom it was made in 1370. Although it is not clear who invented it, the recipe is said to have been so effective in preserving her beauty that the King of Poland asked her, at the age of 72, to marry him (*see Chapter Six for the formula*).

One of the first English perfumes was called *Ess. Bouquet*, created in 1711. This was followed by *Wood Violet*, introduced in 1832. About this time *English Lavender* became popular for the bath and was also used on the handkerchief. It has been a popular scent ever since. *Honey Water* was also a favourite in the eighteenth century. George Wilson, apothecary to James II, claimed that it was good for the skin and had a very pleasant odour. The ingredients for honey water were essence of musk, cloves, coriander, vanilla, benzoin, orange-flower water, alcohol and water.

Pierre-François Lubin created *Eau de Lubin* which was a popular fragrance of the nineteenth century. It included the essential features of an Eau de Cologne with the addition of tinctures of benzoin and Tolu balsams and flower oils. In 1808 Lubin received permission from Princess Borghese to use her name for a fragrance he created which also became very popular.

THE ARRIVAL OF SYNTHETICS

One of the first perfumers to use synthetic fragrances, the making of perfume on the molecular level, was Paul Parquet. In 1896 he created *Le Parfum Idéal* which is still considered a model of perfume composition. Synthetic coumarin and musk ketone were among a number of synthetic components used. Musk ketone was devised by Albert Baur in the nineteenth century. Used as a fixative, it was considered to be the sweetest of the artificial musk scents created. Marcel Billot, a perfumer and founder of the *d'Honneur de Societé Technique des Parfumeurs de France*, said that 'Parquet's great merit is that he was at once able to make use of synthetics in compositions, while nevertheless retaining the homogeneity and fineness provided by natural products'.

DISTILLATION PROCESSES

Although alcohol was not mixed with perfumes for use on the body until the advent of Eau de Cologne, its usefulness in perfume processing was recognized as early as the fourteenth century. The value of this discovery was major, as it provided a useful substance in which aromatics could be dissolved without changing the essential character of their odour. At the same time the distinction between different strengths of alcohols was recognized and this had an effect on the way oils were dealt with.

A number of important books were written on the subject, including *The Noble Art of Surgery and Distillation*, an English translation of a book based on the distillation methods of the Arabians. In Italy various observations were made concerning the difference between the fatty oils which were obtained by expression and oils separated by distillation. Some oils, like aniseed, would congeal, while oils from cinnamon and clove sank in the water. Prepared waters gradually lost their importance and were replaced by the more diverse, fragrant and long-lasting essential oils.

There were six processes used for extracting essential oils from plants at the turn of the nineteenth century. The process used depended on the particular plant and the desired end product.

WATER DISTILLATION

This process is used primarily for plants, barks, woods and a few flowers. The plant material is placed in a still and covered with water. The still is heated until the water boils and the resulting steam passes through a pipe which is surrounded by cold water, causing the vapour to condense prior to reaching the spout. This liquid is allowed to stand until the oil and water separate. The oil is collected and the water is reused and may be bottled as a floral water.

STEAM DISTILLATION

This process is now the most common method of obtaining essential oils. An internal pressure is generated when steam is fed into a vessel holding aromatic plant material, from a vessel which contains water that is being boiled. When the pressure of the material and water become equal to or greater than the pressure of the atmosphere, the mixture will boil. The heat from the steam forces open the cells of the plant containing the essential oil. The resulting vapour is fed through a pipe to a condenser where it is cooled with cold water. The oil will either float on the surface where it can be easily collected, or it sinks to the bottom, in which case the water is carefully drained away, leaving the oil behind.

EXPRESSION

This method is used primarily for citrus fruits. Bergamot fruit, for example, is scarified with fork-like implements which split open the oil glands. The oil is then mopped up with sponges. Rotating drums with many tiny needles were invented early on and the fruit tumbled around inside, eventually releasing all the available oil which would then be collected. Another variation of this method involves rubbing the fruit rind against a grated funnel or putting it in a cloth bag or into a press to be squeezed, thereby releasing the oil.

MACERATION

This process was used for non-delicate flowers such as rose, orange and cassia. Water and fat (for pomade, purified beef or deer suet was generally used while olive oil was used for perfume) are placed in a pan so that the water becomes greasy. The flowers are added and stirred frequently, allowing the fat to draw the essential oil from the flowers. The fat is taken out and strained, fresh flowers are added and the process continues until the pomade is of the desired strength. There are classifications of strengths to which numbers are given: the higher the number, the stronger the fragrance.

ABSORPTION (ENFLEURAGE)

This process was used primarily for delicate flowers such as jasmine and tuberose, whose fragrance would be spoiled by heat. The French felt that this method yielded the finest essence so used it for most of their fine pomades. This process underwent several stages of refinement. Initially, squares of glass were covered with a layer of purified fat and the fresh flowers placed on top. Each morning the flowers were replaced with freshly picked flowers and the process went on during the whole time a particular flower was in bloom. A refinement to this process was introduced which consisted of spreading the flowers on a fine net which was mounted on a separate frame. The net was then slid in between two glass frames which were covered with the fat. The glass and net frames were enclosed in an air-tight recess. Each day the nets would be taken out and the flowers replaced with fresh ones. By using the net, time was saved from having to pick the flowers off the glass.

SOLVENT EXTRACTION

This process is used for delicate flowers. The flowers are put into large tanks with a solvent (ususally hexane), which dissolves the fragrant material along with the colours and the waxes. The solvent is then recovered and a coloured, waxy, heavily fragrant material remains. This material is known as a *concrete* and can be treated by other solvents leaving a coloured fragrant oil known as an *absolute*.

A Parisian perfumer L. T. Piver was credited with having invented another system of absorption which involved the use of a number of perforated plates on which flowers were placed with alternating sheets of greased glass. These were put in a chamber and a current of air passed over the plates to ensure that the scent of the flowers became fixed in the grease.

THE GREAT PERFUME CENTRES

At the turn of this century, Cannes, Nice and Grasse were the principal perfume-making locations, due in part to their climate and soil.

The great expansion of the perfume industry during the latter half of the nineteenth century is due in no small part to the production of synthetics and isolates which came about as a result of advances in organic chemistry. In a short time organic chemistry came very close to imitating natural fragrances. Around the same time new techniques were discovered which simplified the distillation process of plant materials. In 1890 perfumer Joseph Robert established a process for obtaining absolute flower oils which revolutionized the perfumery industry of Grasse.

SCIENCE JOINS ART

The first work with chemistry was by Antoine de Chiris and Roure Bertrand Fils, who presented essences extracted with volatile solvents at the 1900 World's Fair, for which they won the World's Fair grand prize. However it was the French and German manufacturers Schimmel and Haarman and Reimer (H&R) who first reproduced scents of plants and fruits synthetically. One example is benzyl acetate which has a scent like a mix of jasmine and banana. At the time, in 1855, the substance was considered to be an artificial fragrance; however, by the turn of the century Hesse and Müller identified it as a

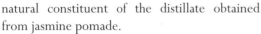

natural constituent of the distillate obtained from jasmine pomade.

Organic chemistry made it possible to determine the chemical constituents of any given plant, which ultimately led to components which contributed to scent being isolated and reproduced synthetically. Indeed, the early 1900s could be considered the 'golden age of perfumery' with organic chemistry providing new perfume bases almost on a daily basis. Chemistry became an important part of the commercial perfume trade.

According to chemist R. M. Gattefossé, who studied the effects of fragrance, there are four classes of synthetic perfumes, the first two being derivatives of natural products:

1. Those that can replace identically the natural product both physically and chemically (for example, geraniol from geranium). These are directly extracted from natural oils without any chemical transformation. The constituents can be separated out easily once their respective points of distillation are known.

2. Those which are different from the original perfume but able to replace it perfectly. These are also directly extracted from natural oils, but have undergone a chemical transformation as a result of the process.

3. Purely chemical perfumes that either imitate a known component or produce a completely new note not known naturally.

4. Those that consist of compound artificial perfumes or fragrances derived from all of the other groups. Natural oils are sometimes included in compound fragrances.

Ready-to-use perfumes, as well as traditional scents, were on offer from a number of perfumers at the turn of the century. They were used to perfume handkerchiefs, pockets, lace, gloves, sachets, but not the skin directly. Synthetic aroma chemicals were also used even though their notes were not as pleasant as the natural sources. But the way was clearly opened for the synthetic fragrance market, marking the turning point between the simple scents of the past and the abstract blends of the future with only the imagination providing the limits.

By 1950 perfume was a mass-market product and more perfume bases were being created to ease the work of the perfumer and, of course, more synthetic fragrances were being used.

HEADSPACE ANALYSIS

In the late 1970s a new technique called *Headspace Analysis* was developed which made it possible to capture and analyse the fragrance given off by a living plant, and to recreate it in the laboratory. Of course it is not 100 per cent accurate, as nature still has many secrets; the odour given off by a flower can change from day to day, and from hour to hour, and individual flowers from the same species may vary. The number of varieties within a species can be enormous and it will take a long time to determine the subtle differences in aroma. While Headspace Analysis techniques are exciting, they are still in their infancy.

One thing has emerged from all this interest in natural smells and that is the keenness of the perfume industry to evaluate and use these new odours to produce even more interesting perfumes based on 'naturals' such as orchids and other exotic flowers. Many fragrant flowers, especially those that do not produce oils in any quantity, are being studied.

The evolution has been brought about principally by Braja Mookherjee, a scientist from International Flavors and Fragrances in New Jersey, USA, who has a deep love and appreciation of nature. Though Headspace Analysis was pioneered some fifteen years ago, it is only in the past few years that it has emerged as a major force in perfumery. Like all great discoveries, the principle is relatively simple: an odour trap (usually a specially designed glass vessel) is placed over a living flower, leaf or other part of a plant; the air is sucked out of the odour trap by a vacuum pump which is connected to an analyser. The air coming out is then carefully analysed and the different components can be identified. Now the actual fragrance of any living flower can be examined. It raises two important issues for perfumery: one is the ability to reproduce identically the actual smell of the flower; the other is the ecological implication of not having to rely on limited natural resources.

It has been known for some time that the oil or absolute from flowers is not always the same. It varies from the time of picking, the species or variety, the time left to dry and so on. There are major differences between the constituents of living and picked (dead) flowers, and these differences are quite marked. Of greater significance is the fact that this method can be used for the tiniest plant or flower. This exciting prospect opens up the whole world of nature to examination. One has only to think of the millions of fragrances produced by the exotic flora of Brazil, Malaysia and other tropical regions to realize what a vista of exploration is now possible.

Perhaps the most interesting area of investigation is the fragrance of the rose. The perfume industry has had to rely on either rose otto (rose oil) which is steam distilled, or rose absolute which is solvent extracted. Now, with this new technology, the actual molecular pattern and structure of the odour of the living rose can be reproduced. Perhaps more significantly, in addition to the roses grown for their perfumery products – *Rosa damascena* and *Rosa centifolia* – other roses can now be examined for their odour quality. Of particular interest, too, is the whole range of rare and expensive florals such as lily of

Lily of the valley, freesia and stephanotis

the valley, freesia, osmanthus, stephanotis and many others. Synthetic perfumes now play a large part in the manufacture of modern perfumes, helping to make perfume products more readily available to the large, and constantly expanding world market.

Chapter Three

CREATING PERFUMES AND FRAGRANCES

A perfume? It is a matter of patience.
JEAN-JACQUES GUERLAIN

Like a painter who is able to mix various paints from his palette to recreate the colours of the setting sun, the perfumer mixes a multitude of fragrances to capture a scent reminiscent of a mood, time or theme. To do this well, the perfumer must possess some essential and seemingly contradictory qualities. A good perfumer must be sensitive to feelings and emotions, since both these elements are heavily interwoven with scent preference. It has become very important for perfumers to have a working knowledge of chemistry as well as an understanding of the production processes involved in making a quality perfume time after time. But the most important characteristics for a perfumer are a fair amount of artistic talent, sensitivity, a good sense of harmony and, probably most important of all, an understanding of people. These qualities characterize the best perfumers, past and present.

THE PERFUMERS

Very little is known about the people who actually prepared the perfume mixtures of ancient times, except that perfume was used mainly for religious purposes; one can surmise, then, that much of this work was carried out by priests. Even as late as the sixteenth century, there were no dedicated perfumers, but people who combined the duties of barber, hairdresser and bathing attendant along with that of perfumer. Valets to the wealthy often performed the duties of perfumer as well. In the early nineteenth century changes in the perfumery trade occurred which transformed it from a craft to the makings of a major industry. Grasse, in France, became a key focal point. It became necessary at that point for perfumers to be trained.

TRAINING

Although the formal training period takes about six years, the most important qualification for a perfumer is to have a good nose. This means that he or she should be able to distinguish a pure product from one that is adulterated. Highly trained noses can actually detect geographical location by smelling a single oil. Olfactory training, which develops olfactory memory, is of utmost importance to the would-be perfumer who will need to be able to identify some 2,000 aromatic substances, along with the blending range of each one. The perfumer must also be aware of the composition of each of the raw materials used so that he or she can judge what the effect each substance will have, how they will affect each other and how they will work together as a whole. These analysis procedures and blending techniques must be learned for the broad spectrum of perfuming which includes cosmetics, body and hair care, household cleaners, room fragrancers – literally everything for which fragrance is important. Gradually the perfumer builds up memories of groups of odours and then mixtures until he or she becomes expert, being able to recognize shades of difference and quality in over 10,000.

A student starts by smelling contrasting odours and then odours which belong to a particular family. Another technique is to have the student try to imitate a particular fragance.

In order to learn smells it is sometimes easier to build a picture-bridge of memory so that a picture becomes associated with a particular fragrance. For example, patchouli was used to fragrance Indian shawls, so a word association for patchouli could be a shawl. What is important is that the associations are meaningful for the person making them. Of course the professional perfumer is able to imagine complete fragrance complexes, add to them and rearrange them – all in his or her mind.

Among the first things to be noticed is that odours change over a period of time and that not all odours have the same evaporation rate. The next step is to determine the evaporation rates of various oils. This is done by placing a drop of oil on a smelling strip and recording the date and time. When the odour changes, the date and time are again recorded. This process should continue until the oil has completely evaporated. It is important that the perfumer has a table indicating the rate of evaporation of all the materials that he or she plans to use in making perfumes.

Jean Carles, a well-known Parisian perfumer, said 'a good perfumer should be able to actually "smell" his perfume prior to the actual blending of the formulation, and starts out by writing down in itemized form his complete selection of components'. Knowledge of each evaporation rate is mandatory so that formulas can be developed according to the volatility of the various products used.

THE STRUCTURE OF PERFUME

The idea of classifying fragrance by evaporation rate or notes was first put forward by nineteenth-century perfumer Samuel Piesse in his book *The Art of Perfumery*. He suggested that fragrances could be classified on a scale corresponding to musical notes, implying that there was a similarity between the organs of hearing and smelling. W. A. Poucher, who wrote *Perfumes, Cosmetics and Soaps* (1923), devised a system of classification of odours based on an evaluation of their evaporation rates. Each perfume material was assigned a number on a scale of 1 to 100 depending on the rate of evaporation. The fragrances could then be divided easily into three sections called top, middle and base notes. Top notes were those rated from 1 to 14 with niaouli being the fastest evaporating and lemongrass, mimosa absolute and palmarosa being the slowest evaporating of the top notes. The middle notes ranged from 15 to 69 and included rose otto and storax oil at the fast end and tuberose and jasmine absolutes at the slower end. The base notes ranged from 70 to 100 and

included materials used as fixatives such as ambergris, benzoin and coumarin, which were all rated 100.

THE BASIC STRUCTURE

However high-tech the industry has become, the structure of most perfumes is still based on three 'notes' – top, middle and base. Top notes are very light and volatile. It is the first odour recognized when the perfume is applied to the skin and provides the first impression. The middle note, or modifier, becomes dominant after the top note has faded, often after only a few minutes. The middle note helps to determine the basic character of the perfume and its fragrance can last for hours. It can also serve to modify the fragrance of the base note, which can sometimes give off an unpleasant initial smell. This is a problem with many scents that are of low volatility and high tenacity; however, over subsequent periods of evaporation they usually change into something quite exquisite. The modifier or middle note can be used to change the unpleas-

COMMON TOP, MIDDLE AND BASE NOTES USED IN PERFUMERY		
TOP NOTES	MIDDLE NOTES	BASE NOTES
Apricot, basil, bergamot, blackcurrant, coriander, galbanum, gardenia, hyacinth, juniper, lavender, lemon, mandarin, melon, neroli, orange, peach, petitgrain, pineapple, spearmint.	*Amber, carnation, civet, cyclamen, geranium, heliotrope, honey, jasmine, magnolia, orris, rose, tuberose, vanilla, ylang ylang.*	*Cedarwood, frankincense, labdanum, leather, musk, patchouli, sandalwood.*

ant initial scent of the base note. The base note gives the perfume its depth. This note acts like a raft allowing the middle notes to float, thereby extending their evaporation time (*see chart above for some popular top, middle and base notes*).

When creating a new fragrance, the base notes are usually decided first. The modifiers, or middle notes will be determined next, as sometimes they must be used to subdue the immediate effect of the base notes. The top notes are the last to be selected, partly because these produce the all-important first impression. It is crucial that the proportions selected – the percentage of top, middle and base notes – achieve a *balanced evolution* during evaporation.

It is also important to produce a good *accord* (the combination of scents blended together to create the new fragrance). Once a basic fragrance is produced, other top, middle and base notes can be added to refine the scent.

Base notes often serve as *fixatives* for the perfume. A fixative is a perfume ingredient which helps the other ingredients retain their fragrance when on the skin by prolonging their evapora-

tion rate. Among the synthetic fixatives now commonly used are: musk ketone, coumarin and vanillin. Other fixatives include animalic, balsam and resin base notes. Sometimes accessory products are included in minute amounts, possibly because their scents are very strong. However, their presence in the formula will completely alter it; even trace amounts provide a unique contribution to the entire accord.

CREATING A NEW PERFUME

Making a perfume is both an art and a science to the perfumer. These nasal wizards create new perfumes, choosing from a wide range of literally thousands of scents. There are four possibilities open to the perfumer when developing a new fragrance. He or she might:
• Reproduce a scent from nature.
• Modify an already existing natural scent.
• Combine a number of natural scents to create a new bouquet.
• Devise a new perfume of his or her own creation.
The array of fragrances at a perfumer's disposal

was originally called a *fragrance organ* although now there are so many fragrances that this concept has been replaced by movable shelves housing hundreds of bottles. At the turn of the century a perfumer would sit at a work station which held tiers of bottles and jars, each containing a different fragrance substance. Now, because the number of choices has dramatically increased with the synthetic explosion, the modern perfumer works in a completely enclosed air-conditioned studio. The perfumer decides which fragrances he or she wants, then these are mixed for the perfumer in a separate laboratory.

THE MIXING PROCEDURE

The perfumer will begin to create a *composition* by mixing the varying materials in tiny measurable amounts, using a very fine electronic balance. The formula will start off with the more solid, often crystalline ingredients, then move up to the more viscous elements and finally finish with the addition of the light absolutes or essential oils. At each stage the formulation must be noted and written down, since it is all too easy to create a masterpiece and then forget to record the proportions or the ingredients!

As the perfume is made, it is stirred so that all the crystalline materials are fully dissolved. Once the final ingredient has been added, the mixture must be fully stirred so that the composition is complete. The mixture rests for about three weeks in a cool place, which allows the components to blend together completely and form a *fragrant whole*. This also allows the compound to 'throw out' any crystalline components. After another three weeks, the perfume is ready to be made into a splash-on lotion, perfume, cologne or other liquid. The final product uses ethyl alcohol (grape spirit) as the solvent.

The perfume is matured several weeks before it is bottled for marketing. This *maturing period* is essential for all perfumes, and without it the perfume remains somewhat rough or unpolished. A perfume not only has to have a fine odour, it has to sparkle in the bottle, remain stable and be able to be recreated identically each time it is produced. This is not as easy as it sounds as characteristics of fragrant materials can vary. It is difficult to get identical batches of oils. Natural materials vary; plant varieties, areas of production, times of growth, techniques of distillation or solvent extraction are all variable factors producing varying scents. And finally the *odour quality* of the material and the nasal ability to assess it may vary. All these problems must be met by the expertise of the perfumer.

THE COMMERICAL VIEW

A truly creative perfumer will not necessarily stay within the range of his or her own smell preference. Indeed, fragrance preference differs significantly from one individual to another, so the market determines what will sell. The most important factors then are the clientele's tastes along with the perfumer's ability to create a fragrance that will become popular out of the almost endless possibilities. There are really no absolute rules to follow when determining which materials to use and what percentages. As a result the creation of a successful fragrance can

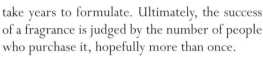

take years to formulate. Ultimately, the success of a fragrance is judged by the number of people who purchase it, hopefully more than once.

The selection of top notes is last on the list although it is perhaps the most important element from a commercial point of view. Most purchasers make their decision based on the first sniff, so the top note has to be the right one, the one that creates that elusive, yet all-important, first impression. The more informed buyer knows, however, that the top note can never be the characteristic note of the perfume; for that, one must wait.

EVALUATING A NEW FRAGRANCE

Perfumer Jean Carles felt that using smelling strips was not the best way to test a fragrance because it did not give the true scent of the perfume. He suggested spraying the perfume for about five seconds into the centre of a room and leaving the room immediately for several minutes, closing the door. The perfume can then be evaluated upon re-entry into the room.

In addition to the true scent being more appropriately detected by vaporization, much time will be saved in the evaluation process because it can take hours for the perfume to fully develop on the smelling strip.

The professional perfumer has a number of other things he or she must consider in addition to the final blend of ingredients that make up the new fragrance. Is the fragrance compatible with the medium it was designed for? Will the final product be possible to produce from the cost point of view; is it a viable economic possibility or are the ingredients too expensive to make it worthwhile? Perfumers must keep abreast of current trends in the marketplace, as well as trends to come; they must also be very aware of what the competition is up to.

SYNTHETIC FRAGRANCES

Hybrids of the stock fragrances produced for the perfume industry are continually being created and improved upon. For example, Folbac (a commercial base) is a jasmine fragrance with tuberose and orange blossom notes which was developed by CAL Pfizer. Charabot, another company dealing with raw material production, has developed a Sambac jasmine which has an orange note added to make the scent more sensuous. A new line of jasmine fragrances called *suprabsolutes*, or 'boosted exotics', has recently been launched by CAL Pfizer. In this process, the standard absolute is improved upon by chemical production techniques.

When developing a floral fragrance, perfumers have found time and again that the best results occur when they keep as close as possible to natural scents, which people find most appealing. Perfumer Poucher claimed that the key to a successful floral perfume was simplicity, using as few aromatics as possible (*see Chapter Four for descriptions of the basic types of fragrance*).

IMITATING NATURE

Synthetic fragrances can be obtained by chemical processes which reproduce the odour of a given plant. The chemicals responsible for the fragrance are isolated from the complex natural mixture of chemicals that make up the entire plant and the chemical structures of those components are analysed. Odoriferous elements in oils that are expensive to procure from one source can sometimes be found in less expensive oils. These elements can be taken from the less expensive oil and, by chemically treating them, a synthetic version of the more expensive oil is produced. For example, synthetic rose oil can be produced from elements taken from geranium and palmarosa oils, making a perfectly acceptable rose oil for perfumery use.

Ionone is a synthetic violet fragrance which is produced by condensing citral (an extract from lemongrass oil) with acetone and treating the product with sulphuric acid. This process was discovered almost by accident, so the story goes. Ferdinand Tiemann and Paul Kruger from the University of Berlin were trying to find the chemical responsible for the violet odour. Rather than using violet oil, which is very expensive, they chose to look for the chemical in orris root, whose odour sufficiently resembled violet for them to surmise that the same chemical was involved. After several attempts which proved fruitless, some strong mineral acids including ionone were poured into the beaker to clean it out, but the addition of that material suddenly produced the desired effect. Ionone, despite initial hesitancy, became a very popular perfume ingredient. Ionone, and a few other synthetic fragrance components that were developed earlier, became the forerunners to the golden age of synthetic organic chemistry.

There are now a vast array of synthetic fragrances available to the perfumer. A number of synthetic fragances including vanillin, coumarin and artificial musks are produced from coal tar. In addition to synthetic fragrances that are imitations of natural ones, many are 'invented' scents, greatly increasing the number of available fragrances. Although synthetics are at a disadvantage to natural products because their odours are usually more chemical and require careful blending in order to produce pleasant odour effects, they do offer consistency in quality and less price fluctuation. Added to this, the fact that synthetic materials are available in practically unlimited quantities makes them a perfect alternative to naturally derived fragrances, particularly now that the use of natural animal products is banned throughout the perfume world.

ENVIRONMENTAL CONSIDERATIONS

In addition to the ever-increasing choice of notes the perfumer has at his or her disposal as a result of synthetic products, there has been such an increase in the demand for fragrances that the supply of natural materials could not satisfy the market entirely. At the same time, the land available for cultivation of plants for perfumery is decreasing as populations grow and land is needed for food crops. Synthetics fill a gap from an environmental point of view as well as from an economic standpoint.

THE GREAT PERFUMERS

There were many great perfumers from the late nineteenth century who made valuable contributions to the art of perfumery. As time went by companies became more protective of what they considered to be trade secrets. Perfume formulas fell into this category and so many of the true creators of popular fragrances remained unknown, shadowy 'backroom' figures. Jean Carles was one of the exceptions. He worked for a company which sold ready-made bases, or 'sauces' as they are known in the industry, to perfume manufacturers who would then bring out a new perfume using the 'sauce' and take credit for the creation. Jean Carles was one of the lucky ones, however, because his efforts were rewarded with the Prestige (de Paris) de la Qualité Française, considered to be the Oscar of the perfume trade. He also won a competition in 1926 sponsored by the *American Perfumer* journal for which a new perfume had to be created.

A keen sense of smell is obviously the most important requirement in the perfume industry. Another distinguished perfumer, Philippe Chuit was gifted with such acuity that he was able to discern purity in the simple substances which he manufactured just by smelling. He could judge the quality of fragrance in the synthetic products produced by his laboratory against the simple chemicals which he knew of in flowers. Chuit created a number of simple synthetic fragrance bases of high quality, several of which have been used in popular perfumes.

Another perfumer with an exceptionally keen sense of smell was Justin Dupont. He remained in the background for the most part, but his ability to detect a new substance which could be used in the creation of a new fragrance was remarkable. He was also responsible for aiding the development of a number of popular perfumes. André Fraysse, Edouard Hache, Paul Johnson, Marius Reboul and Fernand Roudnitska are five other perfumers who have 'anonymously' created many well-known fragrances. No doubt there are dozens of others, but many of the great perfumers have remained in the shadows of their own creations.

THE COPY-CAT MARKET

One of the most interesting aspects of a revolution which occurred in the 1960s was the emergence of a new analytical tool called *liquid gas chromatography*, which was capable of examining the composition of essential oils, absolutes and other compounds. It became possible to examine the composition of oils and identify their basic components, and once this could be done, it was only a matter of time before it was possible to subject perfumes to the same techniques. While it seemed a simple notion, the difficulty lay in separating the individual components represented in the stream of molecules passing up a long fine tube. Once the components had been separated, it was necessary to identify them correctly. This was difficult, for many essential oils could contain as many as several hundred different compounds. As the process developed, more sophisticated methods such as gas chromatography and nuclear magnetic resonance were used.

Perfumes, among other liquids, could be analysed by these techniques so that a similar compound could be made. However, no matter how much information could be found, it was not always possible to duplicate the original in its entirety. But once the previously secret formula became known, it could be mimicked to produce a perfume similar to the original, which is how the copy-cat market was born. Copy-cat perfumes were not usually identical to the original; the more expensive elements which anchored the composition together were often missing so that the perfume was a cheaper version of the original, not only in price, but in ingredients as well. The copy-cat perfume generally did not contain the more expensive 'base notes' or fixatives, and, therefore, had a shorter shelf-life than the original.

Obviously, the original name could not be used, as most perfumes have trade marks to protect them, so, to get around this, the copy-cat version would be described as 'like' or 'similar to' the particular perfume it was copying.

During the 1970s and 1980s, the copy-cat market expanded as the novelty of getting a cheap imitation took hold. Well-known names appeared on market stalls and were sold in 'party plan' operations.

More recently, unscrupulous traders have begun offering the duplicate as the original. Fake perfumes suddenly appeared in many countries and it almost needs an expert to detect the difference in packaging. However, once the perfume is out of the bottle, it can soon be detected as a fake composition, if one is at all familiar with the original. Most people have become aware of this market and know that if you are offered something very much cheaper than the normal selling price, there could be something wrong with the product. It is a case of *caveat emptor* — buyer beware!

MAJOR INGREDIENTS

Perfumes contain a great number of ingredients these days; some are derived from plants, many are now chemically produced. In earlier times materials of animal origin were found to be very useful in perfumery because they held their fragrance longer than any other ingredient. These were used abundantly in the early days of perfume-making and are still very much in use today, but only in the form of synthetic replacements.

Below is an alphabetical list of some of the more common ingredients used in perfumery. Not included are the many natural essential oils; there are numerous books available which give information on all of the major essential oils which are used in perfumery.

Absolute

An aromatic compound produced by solvent extraction; the most common solvent used is hexane although benzine or benzole can be used.

Accord

A perfume composition which is made up of combinations of various single odours which, when blended, produce a new fragrance effect. There is no limit to the number of ingredients that may be included.

Agrumen Oils

A collective term for essential oils from citrus fruits – orange, lemon, tangerine, grapefruit and bergamot are examples.

Alcohol

Widely used in the perfume industry as a solvent (*see Ethyl Alcohol, page 51*).

Aldehyde

Chemicals made from alcohol and some natural plant materials. Discovered at the end of the nineteenth century, aldehydes are used in manufacturing synthetic materials which provide fragrances for modern perfumes. Aldehydes can only be used in very small doses due to the power of their odour. Ernest Beaux was the first perfumer to use aldehydes in large amounts and caused a revolution in the perfume industry when he used aldehydes in his now-famous creation *Chanel No. 5*.

Ambergris

A wax-like substance produced in the digestive tract of the sperm whale, ambergris was sometimes found in oily grey lumps floating in the sea or was taken directly from the body of the whale. In the raw, ambergris has an unpleasant odour; however, once it has been dissolved in alcohol it becomes very fragrant.

For environmental reasons, only synthetic ambergris is used today. Of the various animal scents, ambergris has the highest odour durability and is used primarily for light floral perfumes.

Beeswax

The wax secreted by bees to make honeycombs which can be used in solid perfumes.

Ben Oil

Used since ancient times, this clear and odourless essential oil is obtained from the seeds of the horseradish tree (*Moringa oleifera*). It is an excellent base for perfumes and is also used in the enfleurage and maceration process, as it does not go rancid and readily combines with other fragrances added to it.

Benzaldehyde

A synthetically produced ingredient which is used as an artificial almond oil.

Benzene

Derived from coal tar, benzene is used in the manufacture of synthetic rose scents as well as other artificial fragrances.

Benzoin

A sweet-smelling oleo-resin with a vanilla-like scent which is an excellent fixative.

Benzyl Acetate

A synthetic material used extensively in perfumery because of its low cost. It has a fruity, refreshing jasmine-like odour and occurs naturally in ylang ylang, hyacinth and jasmine.

Benzyl Alcohol

An alcohol which occurs naturally in a number of essential oils such as jasmine and tuberose, but is also chemically synthesized.

Castoreum

This secretion, found on the abdomen of the castor beaver, was known to the Arabs in the ninth century who used it in pomanders. Its disagreeable odour became very pleasing when highly diluted, and it has been used in more modern perfumes as a fixative, providing a spicy or oriental note. Today only synthetic forms are used. Canadian castor has an odour which is similar to pine resin while Russian castor has a more leather-like odour which could be due to the presence of a substance in birch bark, the staple food of that particular animal.

Castoreum contains a complex mixture of over a hundred different substances. The formation of castoreum is due to the fact that the constituents of the food the beaver eats are deposited in the scent glands unchanged. In other mammals these constituents would be excreted in the urine.

Chypre

One of the earliest known blends, dating back to the seventeenth century, it was originally called *Eau de Chypre* but is still used today. The name is said to be derived from its country of origin – Cyprus. The perfume was probably introduced into Western Europe by returning Crusaders. Originally used as a dry perfume, it was made from benzoin, storax, calaminth, calamus root and coriander. Another seventeenth-century recipe was made up of 56 g (2 oz) damask roses, 28 g (1 oz) red sandalwood, 7 g (¼ oz) aloes wood, 28 g (1 oz) giroffle, 12 grains musk, 8 grains ambergris and 8 grains civet. The ingredients were powdered and mixed together. *The Essence of Chypre*, an eighteenth-century French perfume, contained musk, ambergris, vanilla, tonka bean, orris and rose.

Citral

An aldehyde found in a number of essential oils including lemon, lime and verbena, but is is also produced synthetically. Citral is used to give perfumes a fresh lemon-like odour.

Citronellal

Extracted as an alcohol from citronella oil and used in making synthetic perfumes which imitate the fragrances of lily of the valley, hyacinth, narcissus and sweet pea.

Civet

Discovered by the Arabs around the tenth century, this substance is extracted from anal glands of both male and female civet cats – wild cats native to Ethiopia and other parts of Africa. Although the smell on its own is very strong, when diluted it is highly fragrant, and is said to round out the bouquet of a perfume. Civet is considered to be an excellent fixative which is why it is found in many top-quality perfumes. Civetone, the principal component, and skatole, a highly malodorous material, are now made synthetically.

Composition

The harmoniously adjusted individual components that go into the creation of a perfume. Characteristics of the individual ingredients are combined to create a new fragrance.

Concrete

A semi-solid product obtained through the process of extraction using volatile solvents. Stearoptene, an odourless waxy substance, gives the product its semi-solid consistency.

Coumarin

A constituent of many fragrant herbs occurring as a white crystalline substance at the point when the plant begins to wither. The odour is like that of newly mown hay and it is used as a fixative. Tonka bean is rich in coumarin and is used both for its fragrance and coumarin content. Most of the coumarin used in perfumery today is manufactured from coal tar. Coumarin was first synthesized in 1868 by British chemist William Perkin.

Essential Oils

Oily, volatile liquids obtained from plants mainly through a process of steam distillation; used in perfumery for their fragrance.

Ethyl Alcohol

Clear, odourless solvent widely used in perfumery. It is made by the fermentation of starch, sugar and other carbohydrates.

Ethyl Cinnamate

A natural component of storax, it is artificially manufactured and has an oriental scent which is used in amber colognes.

Eugenol

A principal constituent of clove oil but also found in other essential oils such as cinnamon, patchouli and pimento. It is used in synthetic perfumes to give a spicy scent.

Extract

An essential oil mixed with alcohol, which acts as a kind of preservative.

Farnesol

A constituent of musk, farnesol is now manufactured synthetically and provides a much admired lily-of-the-valley scent.

Fixative

Material that slows the rate of evaporation of the more volatile substances in a blended mixture of aromatics while contributing to the overall fragrance of the perfume. Fixatives bind together the other ingredients of the composition so that there is a smoothness of transition from the point when the top notes fade away and the middle notes become dominant and so on.

Heliotropin

A chemically produced aldehyde with the fragrance of heliotrope flowers.

Ionone

A synthetic perfume ingredient made from citral which has a fragrance similar to that of violets.

Isoeugenol

Obtained from eugenol and found naturally in nutmeg and ylang ylang. Imitates the fragrance of carnation. Isoeugenol is also used as a fixative.

Isolates

Obtained from natural materials, such as essential oils, by means of separation.

Linalol

An alcohol found in some essential oils including petitgrain, coriander and lavender. It has a spicy-floral fragrance and is used in perfumes with honeysuckle, lilac or lily scents.

Muscone

A chemical which comes from musk, but is also made synthetically. It is a powerful fixative.

Musk

A powerful fragrance, this ingredient comes from the musk glands found in the abdomen of the male musk deer, indigenous to the high mountainous regions of the Himalayas. The musk mallow (*Hibiscus abelmoschus*) is an evergreen shrub grown in tropical and subtropical climates; when steam distilled, its seeds produce an oil known as ambrette oil which has a musk-like fragrance. The odoriferous element is called muscone and it makes up two per cent of the seed from which it comes.

Nuancer

Fragrance used to support the main ingredients in a composition or to add a special effect.

Otto

The name sometimes given to a distilled oil. Essential oils were called ottos, particularly before the 1900s. The term is still used when referring to rose oil.

Pinene

Although a chemical component of many essential oils, this ingredient is obtained mainly from turpentine oil. Used to make synthetic camphor, citronellal, eraniol and nerol and has a harsh, spicy odour.

Resin

Solid or semi-solid organic plant secretions. They go through a cleaning process before use in perfumery (*see Resinoid below*).

Resinoid

A resin which has been washed with benzene or alcohol to remove the sticky soluble materials.

Reuniol

A combination of geraniol and citronellal which is used as a base for making rose fragrances.

Semi-synthetic Aroma Chemicals

These are synthesized from isolates contained in plants which have been extracted and chemically combined.

Solvents

Odourless, colourless liquids used in perfumery for the dilution of perfume oils. The most commonly used solvent in perfumery production is ethyl alcohol.

Storax

Obtained by expression from the inner bark of the liquidambar tree (*Liquidambar orientalis*) but it is now mostly obtained through chemical synthesization. It is a balsamic oleo-resin and is used in perfume as an alternative to vanilla, ambergris and benzoin.

Styrene

A chemical constituent of several natural balsams, notably storax, it is now synthetically prepared and used as a fixative in floral perfumes.

Synthetic Aroma Chemicals

These are synthesized from basic organic compounds such as coal tar and other carbon products, and are predominantly modelled after natural chemicals, having an identical chemical structure. Some non-occurring natural fragrances are also produced which are new structures resembling natural odour types.

Terpineol

Naturally occurring in a number of plants, but it is mostly used in creating synthetic fragrances such as lilac, lavender, jasmine and eucalyptus.

Tincture

Cold-processed alcoholic extract from natural products.

Tragacanth

An odourless gum exuding from the stems of various species of the astragalus shrub. It is used as an emulsion stabilizer and for binding aromatic powders into pastes.

Vanillin

A component of the vanilla plant (*Vanilla planifolia A.*) which belongs to the orchid family. It is also found in benzoin, balsam of Tolu and balsam of Peru. It is an odourless crystalline substance which appears on the surface of the vanilla bean, making chemical analysis easy. It is a valuable fixative, although today synthetic vanillin produced from a number of sources, including pinewood sap, is most commonly used. Vanillin was first synthesized by Reimer in 1876.

Chapter Four

CLASSIFYING FRAGRANCES

Odours in and of themselves make myths possible.

GASTON BACHELARD

In the same way that fragrance preferences vary from person to person, so too will descriptions of fragrances vary from one person to the next. This holds true for perfumes, some of which are made up of a combination of over a hundred separate fragrances. To maintain some sort of organization, the best that can be done is to group perfumes into types of families which bear a resemblance to one another. This has become increasingly important as the number of fragrances available expands. It is impor-tant to remember that any grouping can only be a guideline; keep in mind that certain perfumes consist of combinations of many fragrances, and therefore they may sometimes fit into more than one group. The choice of just one group would be a matter of opinion.

There is value in classifica-tion for simplicity's sake, as long as it is borne in mind that this also has its limitations. Each perfume family will con-tain a few classic fragrances, some of which helped to initi-ate the family in the first place.

DIFFERENT SYSTEMS OF CLASSIFICATION

A number of attempts at classification over the years has produced several different systems. Indeed even today there is no one classification that is recognized by every perfumer unanimously, although there are a few that are more commonly used either in total or with some variation, and even these get revised periodically. The overriding similarity between all the systems is that they all rely on nature for the category names and basic definitions.

The classification of perfume has undergone a change over the past twenty years, especially as new fragrant materials are discovered from the world of nature and also from the research departments of the large fragrance houses. The impetus for these new materials may be due to the demand for new odours which can blend with existing compounds. There is also an increased demand for perfumes as people in the emerging countries begin to get a taste for them.

EARLY CLASSIFICATIONS

The classification of fragrances was fairly simple prior to the introduction of synthetics. The number of fragrances now available, plus the multitude of combinations, require continual changes to the categories or families of perfume fragrances. It is like adding new shades on to an ever-growing palette. In *The Book of Perfumes* (1865), Eugène Rimmel grouped the various materials that were used in perfumery at that time into twelve categories which were divided according to their nature, as outlined below:

ANIMAL: musk, civet and ambergris.

FLORAL: all flowers available for perfumery purposes, mainly jasmine, rose, orange, tuberose, cassia, violet, jonquil and narcissus.

HERBAL: essential oils from aromatic plants such as lavender, peppermint, rosemary, thyme, marjoram, geranium, patchouli and wintergreen.

ANDROPOGON: essential oils from aromatic grasses, citronella, ginger grass and lemongrass.

CITRINE: essential oils from the fruits of bergamot, orange, lemon and lime.

SPICY: cassia, cinnamon, cloves, nutmeg and pimento.

LIGNEOUS: sandalwood, rosewood, rhodium, cedarwood and sassafras.

RADICAL: orris root and vetivert.

SEMINAL: essential oils from umbelliferous plants including aniseed, dill, fennel and caraway with aromatic seeds.

BALMY: resins from trees, including balsams of Peru and Tolu, benzoin, styrax and myrrh.

FRUITY: essential oils from bitter almonds, tonquin beans and vanilla.

ARTIFICIAL: various flavours produced by chemical combinations. When Rimmel devised these classifications, the most commonly used artificial fragrance was mirbane (nitrobenzene) which had the fragrance of almonds.

Eugène Rimmel also made the first comprehensive attempt to place perfume fragrances into categories in the latter part of the nineteenth century. He suggested that there were eighteen representative types of fragrances which were later extended by some and criticized by others because the system had no scientific basis. His fragrance classifications were: Almondy, Amber, Anise, Balsamic (vanilla), Camphoraceous, Caryophyllaceous (clove), Citrine, Fruity, Jasmine, Lavender, Minty, Musky, Orange flower, Rosaceous, Sandal, Spicy, Tuberose and Violet.

Piesse felt that scents influenced the olfactory nerves in specific ways, making an analogy to the influence of music. Certain odours, when blended together, were like octaves in that they each contributed a different degree of a similar impression, for example, the citrus fragrances of lemon, mandarin and orange. If a blend of different scents was to be harmonious, the fragrances needed to 'chord together'. Although this system was not widely used for classifications, the concept of notes does play a major role in perfume creation (*see Chapter Three*).

Chemists Crocker and Henderson classified odours numerically based on just four sensations which they felt covered all fragrances:

FRAGRANT: derived from the sweetness in flowers, musk and violet ketones.

ACID: the component giving sharpness to an odour recognized as vinegary or sour.

BURNT: carbonized odour.

CAPRYLIC: defined as a generally unpleasant sensation when strong, yet necessary for the overall effect in small amounts.

The system of classification used by pharmacist and perfume consultant William Poucher was to group the synthetics together with essential oils based on their comparative volatility. It took him four years to complete this task which was carried out in a temperature-controlled laboratory, using the best raw materials available and smelling strips, limiting the numbers tested each day to four to protect against olfactory fatigue. For measuring purposes, solid materials were diluted in diethyl phthalate. This classification system, like that of British perfumer Piesse mentioned earlier, was related to the notes concept (*see Chapter Three*).

MODERN CLASSIFICATION

Haarman and Reimer, the fragrance and flavour manufacturers responsible for producing a large number of well-known brand names, produced its first genealogy in 1974 which consisted of approximately 200 feminine perfumes. With all the new fragrances that have appeared since then, the Haarman and Reimer (H&R) genealogy has changed somewhat. In 1986 H&R produced 'a systematic classification of the perfumes on the world market'. The perfumes were categorized into seven main families which were then sub-divided to produce a spectrum of thirteen typical directions (indicated by different-coloured fields). Not all families had

sub-divisions. The colours were meant to optically reinforce the fragrance directions. The families and their sub-divisions are as follows:

FLORAL: floral, floral-sweet, floral-fruity, floral-fresh.

GREEN: fresh, balsamic.

ALDEHYDIC: aldehydic-floral, woody-powdery.

CHYPRE: fresh-mossy-aldehydic, floral-mossy-animalic, mossy-fruity.

ORIENTAL: single division family.

LEATHER: single division family.

FOUGÈRE: single division family.

New fragrances continue to appear. In 1989 H&R produced another version of its genealogy, updating it for the fragrance professional. Its latest genealogy 'was developed upon the premise that every perfume can and must be assessed today against the backdrop of a basic, more or less overriding fragrance concept'. H&R acknowledged that each of these concepts has its own variations. So, rather than classifying fragrances by family or fragrant notes, H&R is now using fragrance 'concepts'. There are three basic fragrance concepts with sub-divisions within each group among the feminine fragrances:

FLORAL: green, fruity, fresh, floral, aldehydic, sweet.

ORIENTAL: sweet, spicy.

CHYPRE: fruity, floral-animalic, floral, fresh, green.

There are five basic fragrance concepts with a number of sub-divisions within each group among the masculine notes. They include the following:

LAVENDER: fresh, spicy.

FOUGÈRE: fresh, floral, woody, sweet.

ORIENTAL: spicy, sweet.

CHYPRE: woody, leathery, coniferous, fresh, green.

CITRUS: floral, fantasy, fresh, green.

As the number of fragrances grows, it becomes more difficult to group them into categories, although there has been a tendency for perfumers to group them in new ways. Trying to fit some into the existing categories does not always work, so a variety of diagrams and charts has emerged. Some new versions put them into a circular or hexagonal form, while another starts from a central core of themes or 'notes' which radiates out encompassing new fragrances which appear. Many of these systems are good explanations of the existing groups while others are seen as working tools.

For the purposes of this book we have chosen to use the major H&R family categories system. However, because the floral family is so vast we have broken it down into its four sub-divisions. Not all perfume companies use one system, but we are doing so purely for simplicity's sake. Also we feel that the placing of a perfume into a category, although based on impressions having a recognizable theme, is subjective. As perfumes become more complex, it becomes increasingly difficult to categorize them.

Because of space constraints we can only provide information on a few selected perfumes. We have tried to pick those perfumes which have been trend-setters when giving broader information, and even then it is not possible to expand on each one. It is important to emphasize that when we do give additional information, it in no way suggests a preference on our part for any particular fragrance. For simplicity's sake, we have chosen to keep to what may be considered a very basic classification. The new fragrances which appear on the market often transcend these barriers, and combinations that were once thought to be entirely contradictory are now appearing on the market. Where a manufacturer has provided a definition that encompasses more than one category, we have placed the fragrance under the category of the first word in the description; for example, Rochas classifies *Madame Rochas* as an aldehydic-floral which we have placed in the aldehydic category. (*Please see pages 104–10 for the lists of perfumes and their classifications.*)

FEMININE FRAGRANCES

The perfume industry is heavily dominated by feminine fragrances, and there are literally thousands of different perfumes in an expanding market. Using the H&R classification we will take a look at this huge group and each of the seven families.

THE FLORAL FAMILY

This is the largest of the perfume families and, as the name implies, it contains fragrances made predominantly from floral scents. Floral perfumes are generally light in fragrance.

One of the common synthetically produced ingredients used in this group is hydroxycitronellal which acts as a fixative and has a sweet odour reminiscent of lily of the valley. Benzyl acetate is also used extensively because of its low cost. Heavier floral notes include rose and jasmine. The floral bouquet prevalent in this fragrance grouping is generally a combination of various white flowers including lily of the valley,

jasmine and tuberose. Floral bouquets made up of a variety of white flowers are sometimes simply called white flower perfumes. Fantasy accords are often included in floral bouquets, although each perfume does have distinct typical notes that distinguish it from another. Floral fragrances are generally used for daytime wear. KEYWORDS: youth, fertility and expectancy.

The floral family contains four sub-divisions:
FLORAL: floral fragrances only.
FLORAL-SWEET: combinations of heavy floral notes with an added sweetness.
FLORAL-FRUITY: agrumen and other fruity constituents usually in the top note.
FLORAL-FRESH: spring-like blossomy freshness.
Some perfumes are classified as floral-oriental, such as *Nuits Indiennes* by Jean Louis Scherrer and *Asja* by Fendi, but for the purposes of our charts have been placed under floral. (*For a list of floral perfumes, please see pages 104–5.*)

THE GREEN FAMILY

This family is directly influenced by the florals and the chypre fragrances and is made up of blends of plant materials such as herbs and mosses in addition to citrus fruits and spicy notes. They tend to be sharper in fragrance and reminiscent of the outdoors. Fresh, floral with oceanic notes, and one of the trends of the 1990s, the green family of scents includes new fragrances such as *Eternity* and *Escape*, both by Calvin Klein, and *L'Eau D'Issey* by Issey Miyake. They are classified here as green florals.

Vent Vert, created by Germaine Sellier of Roure, was the first fragrance of the green family. First introduced in 1945, it remained the only one in its class for close on twenty years. *Vent Vert* was relaunched in 1991 with some fragrance modifications.

KEYWORDS: fresh, outdoors, springtime.

The green family contains two sub-divisions:

FRESH: some perfumers use fresh to denote a cut leave scent, while others use it to describe sharp, citrus notes.

BALSAMIC: mild, sweet, vanilla fragrance.

(For a list of green perfumes, please see page 106.)

THE ALDEHYDIC FAMILY

Discovered at the end of the nineteenth century, aldehydes are chemicals derived from alcohol and some plant materials which form one of the chemical groups known as benzenoid compounds and are used to manufacture synthetic materials for perfumes. Although aldehydes have a sharp, slightly fruity odour on their own, they blend beautifully and sometimes unexpectedly with florals. Some aldehydes can help produce known fragrances while others can create a distinctly new scent. Aldehydic fragrances tend to be much more diffusive than naturals.

Chanel No. 5, developed by Ernest Beaux, was the first aldehydic perfume and has become one of the great classic perfumes. A reporter once asked Marilyn Monroe what she wore to bed, to which she replied '*Chanel No. 5*'.

KEYWORDS: classic, elegant, perfumistic.

The aldehydic family contains two sub-divisions:

ALDEHYDIC-FLORAL: a very important category in modern perfumery, with top notes of aldehydic-citrus or floral notes.

ALDEHYDIC-FLORAL, WOODY-POWDERY: consists of the heavier floral fragrances, citrus or herbaceous top notes with vanilla and woody-powdery notes. Sometimes animalic notes are included.

(For a list of aldehydic perfumes, see page 106.)

THE CHYPRE FAMILY

One of the earliest-known blended perfumes, chypre dates from the time of the Crusades when it was known as *Eau de Chypre*. The name is believed to derive from the island of Cyprus and it is presumed that the perfume was brought to Western Europe by knights returning from the Crusades. Chypre was used as a dry perfume in the seventeenth century; one recipe was made up of benzoin, storax, calaminth, calamus root and coriander mixed together.

The main bases for this perfume family are oakmoss, patchouli, labdanum and clary sage.

Other notes used include lavender, rose, jasmine, bergamot and lemon. The first modern chypre perfume was created by Coty in 1917 and was appropriately named *Chypre*. The main ingredients of that perfume consisted of a top note of bergamot with some lemon, neroli and orange. The floral middle note was primarily made up of rose and jasmine, with the base note having a predominance of oakmoss with a supporting cast of patchouli, labdanum, storax, civet and musk.

Coriandre, created in 1973 by Roure perfumers, has characteristics of the aldehydic family (with a rose and geranium floral middle note) and the chypre family (because of its dry patchouli and moss base). For our purposes we have placed it in the chypre family.
KEYWORDS: elegant and sophisticated.

The chypre family contains three sub-categories:
FRESH-MOSSY-ALDEHYDIC: a predominance of aldehydes and citrus top notes on a jasmine floral heart enhanced by an amber, oakmoss and vetivert base. Typical of this category are *Chant d'Arômes* by Guerlain, *Crêpe de Chine* by Millot and *Ma Griffe* by Carven.
FLORAL-MOSSY-ANIMALIC: a top note of gardenia, hyacinth and galbanum on a heart of rose, jasmine and orris. Enhanced by an amber, patchouli, moss, castoreum, civet and musk base. Typical of this category are *Miss Dior* by Dior, *Koto* by Shiseido and *Empreinte* by Correges.
MOSSY-FRUITY: a predominance of plum and peach top notes on a floral heart of rose and jasmine enhanced by an oakmoss, patchouli and musk base. Typical of this category is *Femme de Rochas* which was relaunched a few years ago to enhance its fruity top note. *(For a list of chypre perfumes, see pages 106–7.)*

THE ORIENTAL FAMILY

This family of fragrances, sometimes also called Amber, contains perfumes which have ingredients of a spicy, woody and exotic nature, reminiscent of the East. It boasts the oldest perfume still available: *Jicky* by Guerlain was created in 1889. Heavy resinous fragrant base notes like musk, sandalwood, vanilla, patchouli, vetivert and benzoin have a distinctive sweetness that forms the character of the oriental perfumes.

Opium, developed in 1977 by Jean Amic, reintroduced the oriental note to both the American and European markets. The top notes are mandarin, plum, clove, pepper and coriander, while the floral middle notes are lily of the valley, rose and jasmine. The base notes consist of labdanum, benzoin, myrrh, opoponax, castoreum, cedarwood and sandalwood. Interesting combinations of fresh and oriental notes, which would have previously been perceived as impossible, are now coming into the market; *Kashaya* by Kenzo is an excellent example. Oriental fragrances are generally used for evening wear.
KEYWORDS: warm, heady, spicy and exotic.
(For a list of oriental perfumes, see page 107.)

THE LEATHER AND TOBACCO FAMILY

These notes, mainly used in masculine fragrances, tend to be fantasy notes containing tar-like, leathery, cured tobacco and animalic elements.

KEYWORDS: pungent, smoky-sweet.
(For a list of leather and tobacco perfumes, see page 107.)

THE FOUGÈRE FAMILY

Fougère, which means 'fern' in French, is a more recent perfume classification. It is used primarily in men's fragrances, which have become increasingly popular. Ferns grow in moist tropical and subtropical regions including the rainforests. There are over 10,000 species of fern.

Fougère fragrances have a lavender emphasis with a herbal quality, particularly the dried herbs and hay odours. In perfumery, the fougère note bears little resemblance to the dominant odour of the actual plants but is a fantasy fragrance made up of synthetic ingredients. It was relatively easy to produce synthetic fougère fragrances using coumarin as a major ingredient. Other notes included in the formulas seem to vary from one perfumer to another. The first fougère fragrance in which coumarin was used was *Fougère Royale* created by Houbigant in 1882.
KEYWORDS: herbal.
(For a list of fougère perfumes, see page 107.)

THE FRAGRANCE CHARTS

To give a very general idea of fragrance classification, two sample charts – one for feminine fragrances and one for masculine fragrances – have been provided in this chapter. It is important to mention that in choosing the fragrances included on these charts the following criteria were followed: all the fragrances are internationally well-known; many of the newer perfume creations were not included as it was felt that these would be less familiar to most people; and last of all, space constraints made it impossible to include anything but a small sample of the vast array of perfumes that are available.

It should also be noted that the number of fragrances chosen in each category is not necessarily related to the importance of that particular family; the choice is designed to reflect as fully as possible the different notes that can be classified into the same broad family. The Appendix provides a full list of fragrances in the categories they are assigned to most frequently (*see pages 104–10*). The listing does not, however, take into account very local brands and unfortunately will become out of date as soon as it is printed.

Finally, it should be noted that the choice of one main descriptive word is an extremely arbitrary way of classifying any fragrance since the mere existence of top, middle and base notes always gives rise to at least three to four main descriptions for any given fragrance.

FEMININE PERFUME FAMILY

NAME	MAKER	Floral	Floral-sweet	Floral-fruity	Floral-fresh	Green	Aldehydic	Chypre	Oriental	Leather and Tobacco	Fougère
Anaïs Anaïs	Cacharel			●							
Alliage	Estée Lauder					●					
Aromatics Elixir	Clinique									●	
Blue Grass	Elizabeth Arden	●									
Cabochard	Grès							●			
Calyx	Prescriptives	●									
Chanel No. 5	Chanel						●				
Chanel No. 19	Chanel					●					
Charlie	Revlon				●						
Coriandre	Coutourier							●			
Diorella	Christian Dior				●						
Eternity	Calvin Klein	●									
First	Van Cleef				●						
Fougère Royale	Houbigant										●
Giorgio	G. Beverly Hills	●									
Joy	Jean Patou	●									
Lauren	Ralph Lauren			●							
L'Origan	Coty		●								
Ma Griffe	Carven							●			
Miss Dior	Christian Dior							●			
Ô de Lancôme	Lancôme			●							
Ombre Rose	Brossard						●				
Opium	Yves St Laurent								●		
Oscar de la Renta	Sanofi		●								
Paloma Picasso	Paloma Picasso							●			
Poison	Christian Dior								●		
Rive Gauche	Yves St Laurent						●				
Shalimar	Guerlain								●		
Vent Vert	Pierre Balmain					●					
Youth Dew	Estée Lauder								●		

MASCULINE FRAGRANCES

Earlier in this century, masculine fragrances were expected to smell of tobacco, leather, fougère or citrus. Today men's perfumes have become less conservative and now include rich, woody, ambery and green notes in their make-up. The *H&R Genealogy of Masculine Notes* consists of twelve notes, representing ten perfume families, as outlined below. A note is a particular scent, whereas a family consists of combinations of notes. The term 'note' can also be used for individual fragrances which influence the total effect of a fragrance. *(For a list of masculine fragrances and their classifications, see pages 108–10.)*

THE GREEN NOTES

These are spicy notes with hints of chypre and floral. *Devin* by Aramis, considered to be a masculine version of *Alliage* produced by Lauder, served as a model for a number of other masculine fragrances using these notes. The principal ingredients in *Devin* are galbanum and bergamot top notes with jasmine, clove and cinnamon middle notes and base notes of cedarwood and frankincense and leather.

THE CITRUS NOTES

This group contains essential oils from the zest of bergamot, lemon, orange, mandarin, tangerine and grapefruit. There are three types of citrus notes: classic, fixed and cool-floral. The classic notes are basically *Eau de Colognes* which have masculine notes. The fixed citrus notes are made up of the fresh sporty colognes, combining citrus with spicy or minty nuances. A good example of this is *Sport Fragrance* by Aigner which has the freshness of citrus combined with lavender and clary sage. *Eau Sauvage*, considered to be a trendsetter, started the category of cool-floral. Created by Edmond Roudnitska, the main ingredients are bergamot, lemon and basil top notes, with jasmine, patchouli and oakmoss.

THE LAVENDER NOTES

This category is usually linked with other fragrance traits belonging to the fougère or woody notes. The spicy-coniferous *Silvestre*-type notes are included in the lavender notes category because of their stimulating herbaceous quality.

THE SPICY NOTES

These are very popular scents in masculine fragrances. *Old Spice*, which contains a citrus top note with middle and base notes of carnation, cinnamon and hints of musk, vanilla and cedarwood, was introduced in 1937 by Shulton and has been a top-selling men's fragrance ever since. A number of fragrances produced in the early 1980s which combined fresh floral with cinnamon spice were well received.

The term 'oriental' was first used in 1978 to describe *Lagerfeld* by Lagerfeld which has a honey-like patchouli note. A number of other fragrances having this oriental touch include *Brisk Spice-Wild Spice* and *Rugger*, both by Avon.

Lagerfeld for Men by Lagerfeld is also classified as oriental. Spicy notes, particularly those with an oriental trend, have continued to be popular. *J.H.L.* by Aramis, for example, is considered to be the masculine version of *Cinnabar* by Lauder which is classified in the oriental family.

THE FLORAL NOTES

These are classified as fresh or dry and are more spicy than feminine florals. *Men's Club* consists of spicy aldehydes in the top note, with rose, jasmine, musk and cedarwood. *Eau de Toilette* was the first of the floral perfumes for men made up primarily of lavender oil. *Claiborne for Men* by Claiborne represents a new trend in floral notes for men which includes ozonic notes.

THE CHYPRE NOTES

These fragrances are made up of fresh, coniferous, animalic-leathery and sweet notes. *Aramis*, produced in 1965, was a trend-setter in the field of animalic-leathery chypre notes. Its main ingredients are a spicy aldehydic top note with jasmine, patchouli, oakmoss and leather. *Halston Z14* by Halston was another trend-setter introducing a fresh-mossy-ambergris scent with bergamot and lemon top notes, and a jasmine, patchouli, amber and leather fragrance.

THE FOUGÈRE NOTES

The herbal, lavender scent with the mossy background characteristic of the fougère note is found in *Brut* by Fabergé, a trend-setter of the 1960s. Its main ingredients are lavender and anise top notes with geranium, oakmoss and vanilla. Another trend-setting perfume in this category was *Paco Rabanne pour Homme* by Paco Rabanne, whose main ingredients consist of a bergamot top note with clary sage, carnation, rosemary, lavender, tree moss and musk.

THE WOODY NOTES

Characteristic of this group are fragrances like sandalwood, cedarwood, patchouli and vetivert with top notes of lavender and citrus. A blend of *Eau de Cologne* and a bitter woody note such as vetivert make a popular modern combination. *Aspen* by Coty is a woody fragrance reminiscent of a pine forest with a citrus top note. A new trend is epitomized by the oceanic notes of *Kenzo pour Homme*, which also has a woody note.

THE LEATHER NOTES

These leathery smells include smokey, burnt wood and tobacco nuances. The fragrance suggests a macho connotation with the leathery scent not necessarily apparent upon initial smell.

THE MUSK NOTES

This became a separate group in the 1970s and the scent is very popular in masculine fragrances. The term is used in the names of many fragrances of which it is a constituent.

The latest trend is the launch of unisex fragrances. Fragrances such as *CK One* by Calvin Klein, a fresh citrus fragrance, is included in this category. No doubt we shall be seeing many more fragrances that will fit into this new unisex classification.

MASCULINE PERFUME FAMILY

NAME	MAKER	Green	Citrus	Lavender	Spicy	Floral	Chypre	Fougère	Woody	Leather	Musk
Antaeus	Chanel									•	
Aramis	Aramis						•				
Armani p. Homme	Georgio Armani		•								
Azzaro p. Homme	Azzaro							•			
Blue Stratos	Shulton		•								
Brut	Fabergé							•			
Cool Water	Davidoff					•					
Denim	Elida Gibbs						•				
Devin	Aramis	•									
Drakkar Noir	Guy Laroche						•				
Eau Sauvage	Christian Dior		•								
Escape for Men	Calvin Klein					•					
Fahrenheit	Christian Dior	•									
Gentleman	Givenchy								•		
Habit Rouge	Guerlain				•						
Jazz	Yves St Laurent				•						
Kenzo p. Homme	Kenzo								•		
Kouros	Yves St Laurent							•			
Macassar	Rochas								•		
Musk for Men	Avon										•
Old Spice	Shulton				•						
Paco Rabanne	Paco Rabanne							•			
Pino Silvestre	Vidal			•							
Polo	Warner						•				
Quorum	Puig									•	
Tabac Original	Maurer & Wirtz					•					
Ted	Lapidus									•	
Trussardi Uomo	Trussardi	•									
Vetiver	Roger et Gallet								•		
Wings for Men	G. Beverly Hills							•			

Chapter Five

USING YOUR KIT

*The very highest recommendation of a fragrance is, that when a
female passes by, the odour which proceeds from her may possibly
attract the attention of those even who till then are intent upon
something else ... for the person who carries the perfume about
with her is not the one, after all, who smells it.*

THEOPHRASTUS

Your *Perfume
Kit* has been
designed
to allow you to
look at perfumes
in terms of their
anatomy rather than
the way in which the
final product fits into a particular classification. Perfumes
are composed of many ingredients and some materials are
found in many perfumes within a classification. This means
that the anatomy of a perfume
group can be simplified so that
it is possible to recreate a perfume without using a vast array
of materials.
The Perfume Kit
contains seven
fragrant com-
plexes chosen
because they rep-
resent most of the
basic building blocks of
perfumery. They will enable
you to create a large number of
different perfume combinations. We have also included a
number of formulas for perfumes, which span almost the
entire fragrance range within
modern perfumery, in order
to get you started. After that,
just follow your nose!

THE PERFUME COMPLEXES IN YOUR KIT

There are seven different bottles of perfume complex in your *Perfume Kit*: *Rose Accord*, *Modern Floral*, *Modern Fruits*, *White Flowers*, *Musk*, *Green Chypre* and *Woody* complexes. Each bottle is fitted with a flow-restrictor cap to guard against accidental spillage and to control the amount of fluid dispensed.

Rose Accord has been selected because of the enormous contribution that the rose has played in both ancient and modern perfumery. Its warm, strong, persistent character makes it a central and indispensable theme in many perfumes. *Modern Floral* is a composition of jasmine, neroli, lily of the valley, tuberose, ylang ylang, narcissus, heliotrope, iris, violet and lilac – fragrances found in many modern perfumes.

The selection of *Modern Fruits* reflects the latest trends in contemporary perfumery which include notes of nectarine, mandarin, guava, grapefruit and peach.

Musk represents an aldehydic musk popular in so many fragrances today. These are synthetic replacement materials as the industry no longer uses animal products for environmental reasons. *White Flowers* is also found in many recent creations and reflects jasmine, tuberose, ylang ylang and narcissus scents. *Woody*, a combination of sandalwood, cedarwood and oakmoss notes, finds its way into many compositions, while *Green Chypre* represents the rich, sensual and rather ancient materials found in the Middle East. The notes which make up this complex are galbanum, labdanum, cedarwood, oakmoss and bergamot.

Each of the seven complexes is examined and described more fully below.

ROSE ACCORD

Rose Note

The rose fragrance in *The Perfume Kit* is made up of a fruity, warm, heady tea rose and rose accord. The best rose otto (essential oil extracted by steam distillation) comes from Bulgaria, Turkey and Morocco but roses are also cultivated for their essential oils in France at Grasse. The season for gathering roses in Eastern Europe is normally just thirty days, from the early part of May until early June. The flowers are collected just before they begin to open prior to sunrise, at the point when they are still covered with dew. At this time of day the blossoms contain the maximum amount of essential oil. They are gathered early in the day because the flowers begin to lose their fragrance as the day progresses; by noon about half of the essential oil content will be lost. Each daily harvest must be processed within twenty-four hours of being gathered. The steam-distilled petals produce a low oil yield – it takes about 3,500 kg (7,700 lb) of blossoms to yield about 1 kg (2.2 lb) of rose oil which makes it a very expensive product. There are approximately 750 rose flowers per kilogramme (2.2 lb) which means that many thousands of flowers go into making even a modest amount of oil. However, a single drop of rose essential oil is strong enough to give its characteristic fragrance to several pints of water.

ORIGINS

There are hundreds of varieties of roses in many different hues. Mythology tells us that the original rose was white, but the blood which ran from Venus' foot when she pricked it with a thorn is supposed to have coloured the petals red. There are a number of species which are used in perfumery, including *Rosa centifolia*, which is purplish-red in colour, *Rosa damascena*, which has large red flowers, and *Rosa alba* and *Rosa gallica*. It is *Rosa damascena*, principally cultivated in Bulgaria, which is the most celebrated for perfumery use. *Rosa damascena* is not a naturally occurring rose, but a hybrid of *Rosa gallica* and *Rosa canina*. There are many rose species other than those mentioned above which are also used in perfumery, each possessing its own unique scent.

As a rule red roses are more odoriferous than those of other colours. William Poucher made the observation that 'roses cultivated in a hot climate have a more powerful perfume...flowers grown under glass develop a finer aroma than those which thrive in the open air'.

The rose is known as 'the queen of flowers' for its fragrance and beauty. It has been esteemed for its perfume from the earliest of times in many different cultures. During the later Tang dynasty in China, the emperor was said to have had a miniature city and gardens laid out in one of the palace halls which was named *Ling Fang Kuo* (Country of Numinous Fragrances). Along with the aromatic trees and ramparts of frankincense, were lakes and rivers of storax and rosewater. A recipe for pressing juice from rose petals exists in China today, which is thought to be quite old. The petals are ground to a paste with water, then the excess water is filtered off and sugar is added to form a jam-like consistency. This can then be stored in porcelain pots and portions diluted when required to provide a fragrant solution.

Otto of roses, which has been known in India since ancient times, is said to have been discovered by Noorjeehan Begum, for whom the Taj Mahal was built. She noticed oily particles floating on the surface of the canal of rosewater which ran through her garden. The scent from the particles was so intensely beautiful that a way was found to collect it by soaking up the oil; rose otto has remained an important perfumery ingredient ever since.

In the ninth century the Arabs had discovered how to produce rosewater by distilling rose petals with water. Large areas in Iraq, Syria and Iran were given over to growing roses for this purpose and rosewater was produced on a commercial scale. It is possible that the art of distillation was introduced into Western countries by Arab traders, and by the tenth century Spain became the first European culture to use rosewater. Today it is *rose otto* (the essential oil extracted by steam distillation) which is used in fine perfumery, most of which comes from Bulgaria. *Rose absolute*, however (obtained by means of volatile solvents), actually captures the truer fragrance of the flower. Most rose absolute comes from Morocco and France.

TRADITIONAL USES

In the first century AD, Pliny listed thirty-two remedies for which rose petals and leaves were used. In folk medicine the rose was known for its antiseptic qualities, making it a valuable cure for a number of ailments. The fragrance of rose was said to aid sleep and so wealthy Romans would fill their beds with rose petals in an attempt to guarantee restful repose.

MAIN CONSTITUENTS

There are six main constituents of rose oil, each providing a different rose-like scent. They are geraniol, rhodinol, citronellal, nerol, linalol and phenylethyl alcohol. Other constituents include eugenol, stearoptene, farnesol and citral.

USES IN MODERN PERFUMERY

François Spoturno, who adopted the name Coty, is regarded as the first of the great perfumers of modern times. He had a very keen sense of smell and was able to detect the presence of an ionone note in the rose capable of giving warmth to a composition. In 1902 he created the perfume *La Rose Jacqueminot* which became an immediate success when a bottle being delivered to a Paris department store shattered on the floor, filling the store with its wonderful aroma. Coty felt that perfumes should be packaged in beautiful bottles, so *La Rose Jacqueminot* was sold in a Baccarat bottle, as were many other perfumes.

MODERN FLORAL

Fruity, peachy, sweet, floral describes the *Modern Floral* complex contained within *The Perfume Kit*. This complex includes notes of violet, lily of the valley, heliotrope, lilac and iris with hints of ylang ylang, peach and rose. The major ingredients are described more fully in the following pages.

Modern Floral bouquet

Heliotrope
(Heliotropium peruvianum or *H. grandiforum)*
Mentioned by Theophrastus and Ovid who thought that the heliotrope's flowering time was dependent upon heavenly bodies. The tiny plant is favoured by garden centres because of its beautiful fragrance and the fact that the purple flowers, which radiate in clusters, bloom for a long time. It grows in tropical and subtropical areas. The odour resembles almonds and vanilla.

Iris

(Iris pallida or I. florentina)

The rhizome of the iris plant when dried produces orris powder which is sold commercially. The powder is a well-known fixative ingredient of good-quality pot-pourri. If the iris rhizome is left for up to three years, it is then used to produce orris butter by solvent extraction. This is a creamy-coloured waxy material which is extremely expensive and has a very fine odour reminiscent of violet. Its unique quality is that it enhances other fragrant materials; as a result, it is used in minute quantities in about one-third of all modern perfumes.

Lily of the Valley

(Convallaria majalis)

There are a number of species of lily and they have been referred to in literature for many centuries. In mythology the lily was sacred to Juno, the Goddess of Light. Different species have differing odours, but can generally be described as delicate and elusive with a honey-like sweetness. Lilies are mentioned in many sacred books, but this term actually describes a variety of flowers which have bell-like flowers, including the Madonna lily, the tiger lily and lily of the valley. The ancient Greeks and Egyptians used the lily to produce a perfume called *Susinon*. The lily of the valley (also known as muguet) is native to Europe and is highly regarded by perfumers who use it in modern fragrances. Both men's and women's fragrances make use of large quantities of muguet. The long curved stem with its white bells is used to produce an absolute.

Lilac

(Syringa vulgaris)

There are numerous species of lilac, but the most well-known is the common lilac (*Syringa vulgaris*). This shrubby tree is a well-known sight in gardens, growing to a height of some 4.5 m (15 ft). The flowers can vary in colour from white to pink and lilac, with the smell differing mainly between the white and coloured types, and the latter having a green character. The lilac has a short flowering period and its flowers can be used to prepare an oil although it has a low yield. Its unique odour resembles a mixture of tuberose, orange flower and almonds. The lilac tree is thought to have originated in Iran (Persia) and was introduced into Spain in the sixteenth century by the Arabs. The flowers were used in pomanders throughout Europe.

Violet

(Viola odorata)

This tiny carpet flower can be found growing naturally in woods and damp places. The flowers and the leaves can both be used in perfumery to produce absolutes; the flower is highly fragrant, but because it is so small the absolute obtained from the blossom is very expensive. The violet leaf absolute is a rich, clear green with a persistent fresh aroma. The violet has long been used for perfumery, medicines, sweets, and love potions. 'Violets, sweet violets' was one of the street calls of London flower-sellers, and the violet is still regarded as a symbol of good fortune. Both the Parma violet and the Victoria violet are used in perfumery.

MODERN FRUITS

Modern Fruits – a tropical, citrus blend

This complex of fruity notes which characterizes many of today's perfumes is created using a number of the fragrant fruity notes which can be found among the citrus oils. *Modern Fruits* is based on the characteristics of nectarine, mandarin, guava, grapefruit and peach, and has a fresh, warm, slightly citrusy aroma which is very reminiscent of these fruits. This new fragrance combination has a special place in very recent perfume compositions, and it acts by accentuating the total note complex.

Mandarin oil and grapefruit oil form the foundation platform on which the complex is built.

Mandarin oil comes in three different forms – yellow, green and red – each having a distinctive fresh, sweet, fruity and warm odour. Mandarins were thought to be the oranges eaten by Chinese mandarins and were also the same colour as the mandarins' robes. Known as the fruit of paradise, refreshing and invigorating grapefruit has a slightly sharp odour which is balanced by the sweet peach, another fruit of Chinese origin, and by guava notes. Guava is an exotic tropical fruit discovered by Columbus in the New World.

Modern Fruits is enhanced by lemon oil and orange oil to produce a truly fruity fragrance which attempts to capture the essence of a group of mainly tropical and exotic plants, a complex which brings new dimensions to modern perfume creation.

WHITE FLOWERS

White Flowers bouquet

The *White Flowers* within your *Perfume Kit* consists primarily of jasmine with a bouquet of other white and creamy blossoms; also included are ylang ylang, tuberose, narcissus, hyacinth and neroli. Jasmine is one of the principal ingredients in modern perfumes; it and the other ingredients in *White Flowers* are outlined here.

Jasmine
(*Jasminum officinale* or *J. grandiflorum*)
A perennial plant with delicate snow-white flowers, jasmine was called 'Moonlight of the Grove' by the Hindu poets. Several species are cultivated for their perfume, including *Jasminum grandiflorum*, *J. officinale* and *J. hirsutum*. Blooms are harvested at night or at dawn when, due to chemical action, the fragrance is strongest. After picking, the petals are left to 'breathe' for a period of time before the oil is extracted. In Egypt,

the flowers are hand-picked one by one, very gently without bruising the flower, by young girls. Bruised flowers deteriorate very rapidly which is one of the reasons why attempts to mechanize the picking process have failed.

The fragrance produced varies depending on when the flower was harvested. The crop picked at the beginning of the season will have different fragrance characteristics from those picked at the end of the picking season, which usually has a span of about three months.

The preferred method of extraction for jasmine is by enfleurage because it provides a relatively high yield of flower oil. Unfortunately, this is a costly procedure due to the amount of time and effort involved.

Today, volatile solvent extraction is the method most commonly used. The flowers are placed in large containers and the solvent (usually hexane) is added. The container rotates, or the solvent circulates, allowing the solvent to extract the oils from the flowers. The oil-enriched solvent is then led to a vacuum still where the solvent is distilled and can then be reused. Several additional stages occur to remove the wax and the alcohol. An estimated 1,000 kg (1 ton) of fresh flowers yields approximately 2.5 kg (5.5 lb) of jasmine concrete, with 1 kg (2.2 lb) of jasmine concrete yielding about 550 g (20 oz) of jasmine absolute.

The concrete is obtained by extraction from the flower of *Jasminum officinale* as well as other species of *Jasminum* (*J. grandiflorum* is sometimes grafted on to *J. officinale* to make the flower more resistant to frost conditions). The consistency of the concrete is a waxy, brownish-red mass with a jasmine odour. The absolute, an important ingredient in perfumery, is derived as a result of stearoptene, the waxy substance, being removed from the concrete with the use of alcohol. The absolute is a yellowish-brown liquid which has a warm floral aroma.

The first commercial production of jasmine for perfumery started in the late nineteenth century in Algeria, with Italy following in the 1920s, and Morocco and Spain in the 1950s. Egypt, where the climate is very suitable, started production in the 1970s, the same time as India. Today Egypt produces more than 60 per cent of the world's jasmine concrete. India produces 22 per cent, with the remainder coming from China, Morocco, Guinea, France and Algeria.

MAIN CONSTITUENTS

The main constituents of jasmine oil are benzyl acetate, linalyl acetate and linalol. Other major components are nerol, nerolidol, indole, phytol, isophytol, methyl linoleate, faenesene, terpineol, benzylalcohol, benzaldehyde, cresol, eugenol, jasmone and acids.

USES IN MODERN PERFUMERY

The first perfume to contain jasmine and aldehyde notes was *Coup de Foudre*, developed by Henri Almeiras for Rosine in 1918. Although the favourite fragrance of that time tended to be violet, a number of other jasmine-scented perfumes were available. Jasmine is the principal ingredient in a number of the great classic perfumes including *Arpège*, *Chanel No. 5* and *Joy*. It is estimated to appear among the principal ingredients in about 83 per cent of all quality perfumes. Jasmine has a very unique odour which is not easy to imitate by synthetic processes.

THE SYNTHETICS

When the first work with artificial fragrances began in 1855, less than ten of the constituents of jasmine had been identified. Other chemicals were soon discovered and by 1933 the perfumer Maurice Chevron had created *Jasmin 231*, made up of benzyl acetate and synthetic aroma chemicals. Benzyl acetate is a colourless liquid with a fruity, jasmine-like odour which can also be manufactured synthetically. It is considered to be one of the most useful synthetic materials. Because benzyl acetate is inexpensive to produce synthetically, it is now a basis for almost all jasmine perfumes. *Jasmin 231* was well-received and is still in use today.

Numerous other bases were developed to ease the work of the perfumer, including *Fleur de Jasmin* which has an orange blossom tonality, and *Floraline Jasmin 62*, a jasmine base with the scent of jasmine blossom and raspberry jam.

One of the world's most widely sold synthetic aroma chemicals is known as hedione. The name comes from *Hediosmia*, an imaginary Jamaican flower mentioned in a book by Joris-Karl Huysman, which supposedly had a scent like French jasmine. Hedione was first introduced in the early 1960s but did not become popular until it was used in small amounts in *Eau Sauvage* by Dior. *Headspace Jasmine* is a base that was derived from new techniques developed in the late 1970s (*see Chapter Three*). The first perfume to contain it was *Eau de Givenchy* created by Daniel Hoffman in 1980. A number of other bases have been created as a direct result of Headspace Analysis. They include *Arabian Jasmine*, *Jasmine Polyanthum*, *Full Jasmine* and *Sambac Jasmine*.

Ylang Ylang
(Cananga odorata)

'Ylang ylang', which means 'flower of flowers', is steam distilled from the flowers of *Cananga odorata*, and has a sweet, jasmine-like fragrance. First introduced in Europe in 1864, it was one of the ingredients used in Macassar oil. It is now an ingredient in many high-quality perfumes.

Tuberose
(Polianthes tuberosa)

This long slender plant with an erect stem bears between ten and fourteen flowers from which the oil is extracted. Its name comes from Latin, *polos* which means 'city' and *anthos* meaning 'flower'. Another name for this plant is night hyacinth, as the flowers emit a delightful intoxicating fragrance reminiscent of a garden of flowers in the evening. It has a heavy, honey-like fragrance. Like jasmine, extraction of tuberose essential oil is by enfleurage. The blossoms are hand-picked, wrapped in damp cloths and processed immediately. It takes 150 kg (330 lb) of flowers to yield 1 kg (2.2 lb) of pomade. The yield of absolute extracted by volatile solvents from the blossoms is so small that it makes tuberose absolute one of the most expensive of all the perfume materials used today.

Hyacinth
(Hyacinthus orientalis)

The hyacinth is native to Syria and other parts of the Middle East. It was described by the epic poet Homer as a bell-shaped flower which formed the couch of Jupiter and Juno; it figures in Greek mythology and was worn in crowns by virgins at their friends' weddings. Ovid attributes its name to Hyacinthus, a beautiful Spartan youth who was accidentally killed by Apollo; the flower grew where his blood fell on the ground. It was introduced into Britain in the sixteenth century and some 4,000 varieties exist. Enormous areas are devoted to the cultivation of these plants in the Netherlands and, when in bloom, the entire area surrounding the fields becomes heavy with the exotic smell.

Narcissus
(Narcissus odorata or N. jonquila)

This flower is associated with beauty and romance as depicted in the myth of Narcissus, the son of a river god and a nymph, who fell in love with his own reflection in a spring. He was so taken with himself that he pined away from too much love of his own beauty. The flower that sprang from his grave was dedicated to him and is also the flower that Persephone gathered in the meadows when Hades, the king of the underworld snatched her away. Pliny states that the narcissus has narcotic properties and its odour is quite heavy when the flowers are smelled in profusion. Though grown principally as a garden plant for florists, it is used in the perfume industry where it is made into an absolute. Its fragrance is a little like a blend of jasmine and hyacinth. The Romans used narcissus flowers to produce a perfume unguent called *Narcissinum*.

Neroli
(Citrus aurantium)

The original orange tree in most of Europe was the bitter orange; it was not until the fifteenth century that the Portuguese introduced the sweet orange from the East. In the sixteenth century the Italian philosopher Porta is reputed to have distilled an oil from the flowers. It was not until 1680 that the oil from the flowers was distilled and called essence of neroli, which the wife of Flavio Orsini, Prince of Neroli, used for perfuming gloves. Another origin of the name 'neroli' is said to be from Anne-Marie de la Tremoille-Noirmoutier, the second wife of the Prince of Nerola. She is said to have used it to perfume *her* gloves. Neroli is now produced in France, Morocco, Tunisia and Egypt, and is used in both men's and women's perfumes.

There are two distinct essential oil fragrances that come from the orange blossom (bigarade or Seville orange) depending on the process used. If the maceration process is used and the resulting pomade is dissolved in rectified spirit, the result is orange-flower oil with a fragrance very close to that of the original flower. However, if the orange blossom is distilled then the resulting otto is known as neroli oil. Orange-flower water is a by-product of the distillation process. A quite separate essential oil is obtained from the zest of the orange and is not to be confused with neroli, obtained from blossom.

MUSK

The *Musk* in your *Perfume Kit* is an aldehydic, musky floral. A valuable perfume and fixative for many centuries, musk was originally introduced into Europe from Arabia, although it was thought to have been first used by Chinese mandarins to perfume their robes. In 1189 musk was one of the gifts given to the emperor of Rome by Sultan Saladin. The male musk deer has an abdominal scent gland which produces the musk as a sex lure. This musk is known as musk tonquin and contains less than 2 per cent of a ketone called muscone, the main cause of the scent. According to research on the subject, there is a residual sensitivity in humans to this musk fragrance and so musk-like scent became the base for many 'pulse twitching' perfumes.

Musk odour is one of the most intense of all known smells. The minutest amount is enough to impart its characteristic odour to a large body of air. On its own the musk fragrance is much too powerful and so it is diluted and mixed with other fragrances, often flowery notes, producing a scent more alluring to the human species. A popular fragrance in modern perfumes, it is one of the best fixatives available.

Although there were a number of attempts to make synthetic musk, the first patent was awarded to Albert Baur from Gispersleben, Germany, who in 1888 produced *Musk Baur*, also known as tonquinol. There are now about twenty different synthetic musks used in perfumery today. Tonquinol is not among them; two other musk synthetics that Baur produced – ambrette and ketone, considered to be the sweetest of the artificial musks – are still being used. Although part of the reason for producing synthetic musk is to protect the animals, it also ensures consistency of odour which is difficult to achieve when using natural materials. (*See page 106 for a list of commercial perfumes which contain a musk base.*)

GREEN CHYPRE

Cedarwood, labdamum and oakmoss

The original chypre was a heady combination of natural materials produced in Cyprus at the time of the Ptolomaic Egyptians. The *Green Chypre* contained in *The Perfume Kit* includes galbanum which produces the 'green' note which is characteristic of many perfumes. Other ingredients in this blend are labdanum, cedarwood, bergamot, oakmoss and amber. The description for amber, which is a constituent of various complexes, is provided under the *Woody* complex (*see page 76*). The main ingredients of *Green Chypre* are outlined on the following page.

Cedarwood

(Juniperus virginiana)

There are many cedars grown in temperate and tropical regions which produce oils used in perfumery. In most cases, the oil is distilled from the chips and shavings which are the by-products of the furniture or timber trade. The odour is strongly woody and pungent, reminiscent of newly sharpened pencils. Other cedar oils may have a different odour though all tend to have strong woody notes. The cedars of Lebanon, mentioned in the Bible and other ancient texts, were well-known for their keeping qualities and were used for Egyptian sarcophagi.

Labdanum

(Cistus)

A sweet-scented oleo-resin which comes from the shrubs of the genus *Cistus*, labdanum is also known as the rock rose, which grows in the Mediterranean. There are a number of species of labdanum used in perfumery. The resin exudes in sticky droplets on the underside of the leaves and stems. The fragrance of labdanum closely resembles ambergris, and as it is economical to use and mixes well with other perfumes, it is used extensively in perfumery.

Oakmoss

(Evernia prunastri)

Used as a fixative by perfumers, oakmoss has an odour which is warm, ambery and somewhat musky. It is derived from lichens which grow on oak trees at high altitudes throughout Europe. It is extracted by hydrocarbon solvents to produce concretes, and by alcohol extraction of concretes to produce absolutes. Resins and resinoids are also produced. Oakmoss is considered to be a superb perfume ingredient in small amounts.

Bergamot Oil

(Citrus bergamia)

Bergamot oil is produced by cold expression from the peel of the fruit of the bergamot tree, which grows almost exclusively in southern Italy. The oil is a green colour with an extremely rich, sweet and fruity fragrance. Bergamot is often used as a top note in perfumery.

WOODY

Sandalwood and Vanilla

The *Woody* complex, with its cedarwood, sandalwood, amber, vanilla and musky notes, forms a base for many perfumes, both masculine and feminine. The impression it gives is of a strong, natural forest-like odour with a cool, fresh, but strong scent. Its composition is firmly based on a forest or woody foundation but made richer and deeper with other ingredients. The main constituents of this *Woody* fragrance are outlined on the following page (*see the Appendix, pages 104–10 for a list of woody perfumes*).

Amber

Usually a shortened form of ambergris, which is no longer used in the perfumery trade, amber is now recreated chemically as an aroma material which closely resembles both the odour and characteristics of the material found in the sperm whale. Amber tends to be used as a fixative and gives a unique quality to perfumes.

Oakmoss
(Evernia prunastri)

Moss is a lichen which grows on the bark of trees. Oakmoss absolute and oakmoss resin are made by extraction and used in perfumery as fragrance fixatives, providing warmth and extraordinary lasting quality to a composition (*see also under* Green Chypre *complex*).

Sandalwood
(Santalum album)

Perhaps one of the most important essential oils used in perfumery and fragrance compounds, sandalwood oil is thick and viscous with a most persistent odour. This sweet, woody oil is produced from the heartwood of the stem and roots of the parasitic tree. The best-quality oil comes from the stills run by the Indian government where it is known as 'Agmarked Sandalwood Oil'. It has long been used in Ayurvedic medicine and is used in the funeral pyres of dignitaries. Sandalwood oil finds its way into a large number of perfumes, both masculine and feminine. Its rich, underlying aroma blends well with most aromatics and gives a richness which is highly appreciated.

Vanilla
(Vanilla planifolia)

The vanilla pod was discovered by the Spaniard Cortés when he invaded and conquered Mexico in the sixteenth century. The Aztecs used it to flavour their *chocalatl*, a cocoa drink. The fruit pods of the orchid vine vanilla are dried and produce long black pods with a sweet vanilla-like odour which is used extensively as a flavouring and as an ingredient in perfumery. Almost a quarter of all modern perfumes use vanilla as an ingredient, though vanillin, the whitish crystalline material found on pods, can now be produced from other plant materials.

CHOOSING A PERFUME THAT IS RIGHT FOR YOU

There are a number of reasons why we decide to purchase a particular perfume which we have not used before. We can be influenced by a big publicity campaign for a fragrance which suddenly has become all the rage. Many people like to feel that they are in tune with the latest fashion. It is easier (and cheaper) to achieve that with a fragrance than it is with clothing.

We sometimes buy new perfume because we have admired the fragrance on someone else. It is important to remember that each of us has a different body chemistry, so any given fragrance may smell quite different on you than it did on the person you may have admired wearing it. Although fragrances come and go out of fashion, some have withstood the test of time, and this too can be another reason for purchasing it.

Whether the perfume you are considering purchasing is the latest fashion or an old classic, the important questions you must ask yourself are 'how does it smell on me?' and 'does it have the desired effect?'. To test these two points, there are a few simple ground rules for evaluating a perfume. First, never judge a perfume by the first sniff as this only gives the top note, which does not linger long. Placing a drop or spraying a small amount on to your wrist may be hampered by the fragrance you put on that morning or indeed the smell from your clothing or even a glove, if you are wearing one. Also,

there may be many other fragrances in the air, particularly if you are at a perfume counter in a shop. The best way to judge a perfume is by wearing it. Most perfumes are now available in sample sizes as small as 1 ml (0.03 fl oz). It makes sense to purchase a small amount first, use it and decide how you feel and what reactions you get. The other option, of course, is to create your own fragrances.

THE PSYCHOLOGY OF FRAGRANCE

It is not by accident that a fragrance sometimes expresses its personality in its name. The ingredients used to create a fragrance are chosen based upon the desired fragrance type to be created. This can sometimes be helpful when trying to buy a perfume for a friend.

Familiar smells are often associated with particular memories; these memories and the emotions they provoke then become associated with that particular fragrance. It is for this reason that a chart of fragrance types, while useful, can only act as a guideline.

Floral fragrances have a fresh smell that suggests robustness and lots of energy. The flower is at its most fragrant and alluring when in full bloom, exuding forth its nectar or plant pheromones. The fragrance it gives forth is that of fertility, expectancy and optimism, some of the qualities of youth. It is no wonder that most of the fragrances aimed at attracting the attention of teenagers have a floral base. Sweet and fruity odours are popular among the pre-puberty age group, but at puberty odour preferences appear to change towards musky, flowery scents.

REFRESHING A 'TIRED' NOSE

If you have been smelling a number of perfumes your nose may get 'tired', even after just six perfumes. One way to attune your nose to smelling a range of perfumes is to bury yourself into your own scent. You simply turn your head into the clothes that you are wearing – a blouse or coat, for example. By taking several deep breaths you will clear out any lingering fragrance particles so that your nose is ready for smelling again.

GETTING STARTED

To start creating your own fragrances, all that you will need, in addition to the complexes provided in your *Perfume Kit*, are some small glass bottles. Obviously, you will need one bottle for each blend that you create. These bottles are available in a variety of sizes (from 3 ml [0.1 fl oz] up) from specialist suppliers (*see page 111*). You may also be able to purchase these bottles from your local chemist or drug store. Further supplies of the perfume complexes are available from Butterbur & Sage (*see page 111*).

GENERAL INSTRUCTIONS

The complexes contained in your kit are in bottles with flow restrictors to enable the liquid to be dispensed by the drop. In one of your empty glass bottles mix the required number of drops from each complex required for that particular formula. Once you have added the required number of drops, replace the lid on the bottle and gently shake your composition in order to blend each of the elements. After you have made a few compositions you should be able to start your own experiments. You could begin to branch out by altering or adding to one or two of the formulas provided on the following pages. Let your own personal fragrance preferences and

your nose be your guide.

All perfumes have a shelf-life. Keep in mind that an unopened bottle of perfume will have a shelf-life of about two years, while an opened bottle will have a shelf-life of about nine months. Seal your own creations well and store them out of the light in order to preserve them.

SOME BASIC HINTS

The more businesslike the occasion, the lighter a perfume should be. Basically, the lighter, fresher and more fruity a fragrance is, the less it is considered to be seductive. For evening wear, a more concentrated perfume is best. In warm weather, fresh, flowery notes are recommended and in winter, stronger chypre or aldehydic notes are preferable. Heavy, warm fragrances give a feeling of closeness and intimacy and so are more appropriate for romantic occasions.

We have provided forty-five formulas in ten categories, from romantic to sophisticated. Once you decide on the image you want to portray, there is a choice of perfume blends in each category. When you become familiar with which combinations are most suitable for particular situations, you can make up your own formulas by varying the number of drops used.

ROMANTIC

The perfumes in this category represent the fresh idealistic person with strong romantic yearnings and leanings. These fragrances would suggest a person who has the expectation of possible attachments lying in wait. The vision is of a home, family and security in the future. Notions about life tend to be warm, cheerful and comforting. These blends are predominantly floral fragrances, with the allure of musk to create an aura of romance.

MUSKY ROSE ACCORD

A heady, aldehydic musky floral with rose notes.
Musk...5 drops
Rose Accord..1 drop

MODERN FLORAL MUSK

A contemporary mélange of white flowers enhanced with musk notes.
Modern Floral...5 drops
Musk...2 drops

MUSKY WHITE FLORAL

An aldehydic, white floral musky composition.
White Flowers..5 drops
Musk...13 drops

MODERN MUSKY FLORAL

A musky bouquet of white flowers refreshed with a delicate peachy note.
Musk...5 drops
Modern Floral...1 drop

SWEET WHITE FLORAL

A radiant harmony of white flowers and modern fruits.
White Flowers..6 drops
Modern Fruits...1 drop
Musk...3 drops

PROFESSIONAL

These blends reflect a smart, more directed and determined approach to the whole concept of perfume. The wearer will use these perfumes as an element of armour in the image that she presents to the outside world. Strong, somewhat severe, with the need to project a more powerful image in the boardroom, bank, business or commercial centre. Fragrances are light, unobtrusive, delicate blends.

GREEN MUSKY CHYPRE

A light, green floral chypre blend.
Green Chypre...10 drops
Musk...7 drops

MODERN WHITE FLORAL

A delicate blend of white flowers with a light, fruity modern top note.

Modern Floral..5 drops
White Flowers...1 drop

WHITE FLORAL ACCORD

A radiant, light blend of flowers, both white and modern, with an exotic woody base.

Modern Fruits.......................................4 drops
White Flowers.......................................13 drops
Woody...2 drops

FRUITY WHITE FLORAL

A very contemporary light, fruity floral blend.

White Flowers...8 drops
Modern Fruits...3 drops

FLORAL FRUIT

A floral fruity blend with a delicate touch of musk.

White Flowers...6 drops
Modern Fruits...2 drops
Musk...2 drops

SPORTS AND LEISURE

Life is an outgoing and ongoing experience to be enjoyed to the full and can encompass many different environments and situations including both participation or observation of sporting events. The person attracted to these perfumes might find herself happily wearing one of these perfumes while playing tennis, working out at the gym or simply enjoying an invigorating walk through peaceful countryside. She might be just as happy to wear it while sitting on the sidelines. The perfumes in this category tend to have fresh green floral notes.

MUSKY GREEN CHYPRE

A green, fresh chypre perfume enhanced with warm, musky undertones.

Musk...5 drops
Green Chypre...2 drops

WHITE FLORAL CHYPRE

A fresh blend of green and floral notes with a dominant chypre character.

Green Chypre...5 drops
White Flowers...2 drops

MODERN FLORAL GREEN CHYPRE

A floral chypre refreshed with green accents, reminiscent of the countryside.

Modern Floral...5 drops
Green Chypre...2 drops

SEDUCTIVE

These are perfumes for the yearning woman. They reflect a warm, responsive, emotionally involved person yet one who is totally in charge of her own destiny and opportunity. These 'signature' perfumes are for the daring and adventurous, being made up of warm, heady oriental fragrances. Just a few drops of one of the following blends behind the ears, on the neck and anywhere else you care to imagine will help set the mood.

FLORAL WOODY ORIENTAL

A floral blend with sweet oriental accents.

Woody...5 drops
White Flowers.......................................1 drop

MUSKY FLORAL ORIENTAL

A floral blend enhanced with musky oriental notes.

Modern Floral......................................4 drops
Green Chypre.......................................3 drops
Musk...8 drops

BALSAMIC ROSE

A soft warm rose with woody accents.

Rose Accord.......................................15 drops
Woody...1 drop
Musk...3 drops

FLORIENTAL CHYPRE

A modern floral with woody musky accents enhanced by a chypre character.

Green Chypre..1 drop
Musk...2 drops
Woody...2 drops
White Flowers.......................................3 drops

MODERN FLORAL ORIENTAL

A heady, warm, woody modern floral blend with enticing oriental tendencies.

Modern Floral.....................................10 drops
Woody...3 drops

YOUTHFUL

Exuberance and *joie de vivre* typify these perfumes. Life is the tracing out of new paths and discoveries. Light, fresh, fruity and eternally young, there are always new doors to be opened and new vistas to enjoy. These fragrances tend to be modern fruity florals with musky hints.

MUSKY MODERN FLORAL

A contemporary floral blend with heady, warm, woody, musk undertones.

Modern Floral......................................5 drops
Woody...3 drops
Musk...7 drops

SWEET FRUITY ROSE

Sparkling, floral composition with a fresh, fruity top.

Rose Accord.................................20 drops
Modern Floral...............................2 drops
Woody..1 drop

MODERN FRUITY FLORAL

A contemporary blend of modern fruits and flowers.

Modern Floral...............................6 drops
Modern Fruits...............................1 drop

FRUITY WOODY FLORAL

A fruity, floral blend with warm, woody undertones.

Modern Floral...............................2 drops
Modern Fruits...............................1 drop
Woody..4 drops

EVERYDAY WEAR

Personality, consolidation and image are the keywords for this group of perfumes. They can be worn around town yet are discreet enough not to shout your presence. Comfortable fragrances, these are the ones which you ought to be able to carry along with you for all situations. Floral and white floral bouquets are popular for this category.

FLORAL BOUQUET

A delicately balanced, contemporary floral blend.

White Flowers...............................2 drops
Rose Accord.................................3 drops
Modern Floral...............................2 drops

FRESH FLORAL

A sparkling, light, rosy floral bouquet.

Green Chypre................................4 drops
White Flowers...............................2 drops
Rose Accord.................................6 drops

WHITE FLORAL MUSK

A radiant floral bouquet with musky undertones.

Musk..5 drops
White Flowers...............................1 drop

FLORAL ROSE BOUQUET

A fresh floral bouquet with added rose rounded off with musk notes.

Rose Accord.................................4 drops
White Flowers...............................1 drop
Musk..3 drops

PARTY

Gregarious and fun-loving, burning up energy at a pace that leaves most people breathless – this is the personality reflected by these fragrances. Daring, energetic and always the centre of attention, this personality generates excitement regardless of the circumstances. The party is the scene, the party is now and the party is you with all your engines roaring on full power. This category is made up of warm, fruity, musky fragrances.

MODERN FRUITY MUSK

An exotic blend of fruits and musky notes.
Musk...10 drops
Modern Fruits......................................1 drop

MUSKY FRUITS ROSE BOUQUET

A fruity blend with rosy floral musk accord.
Rose Accord.......................................4 drops
Musk...2 drops
Modern Fruits......................................1 drop

MODERN WOODY FRUITS

A contemporary composition with fruity overtones and a deep, warm, woody base.
Woody...5 drops
Modern Fruits......................................1 drop

FRUITY ROSE

A lively floral with fruity accents.
Rose Accord......................................10 drops
Modern Fruits......................................1 drop

MODERN FRUITY CHYPRE

An enticing fruity chypre with a fresh green top note.
Green Chypre.....................................10 drops
Modern Fruits......................................1 drop

MUSKY FRUIT

A sparkling contemporary blend of fruits with warm, woody, musky depths.
Modern Fruits......................................1 drop
Woody...4 drops
Musk...6 drops

MATURE

These are the appreciative years when all the apparent fripperies of youth have fallen away. It reflects a personality happy with what life has to offer. Now you can wear and enjoy your perfume for what *you* feel, rather than using it as a tool to bolster an image you want to project to the world. You are your own independent and mature person. Now is the time to enjoy all of life's nuances and mature experiences. The fragrances found within this group are considered to be classic floral combinations with sweet connotations.

MODERN ROSE BOUQUET

A heady floral combination with enhanced rosy notes.
Modern Floral..10 drops
Rose Accord..1 drop

SWEET WOODY FLORAL

A sweet heady floral composition with full-bodied, woody base note.
Woody...5 drops
Modern Floral...1 drop

WOODY ROSE

A delicate floral bouquet with a powdery balsamic woody note.
Woody...5 drops
Rose Accord...2 drops

SOPHISTICATED

You are the sort of person who is at home on the Rue de Rivoli in Paris, and are totally in charge of your destiny. Your hair, make-up, clothes and accessories are totally right, but totally you. Nothing is left to chance. These perfumes reinforce this statement about your image which is chic and smart. They are most often woody chypre blends.

WOODY GREEN CHYPRE

A fresh chypre enhanced by a warm, woody base note.
Woody...5 drops
Green Chypre..2 drops

ROSY GREEN CHYPRE

A lively chypre composition delicately toned down with the added floral and rose notes.
Green Chypre...5 drops
Rose Accord..2 drops

WOODY CHYPRE

A full-bodied green chypre with the added sophistication of woody depths.
Green Chypre...5 drops
Woody...1 drop

MUSKY FRUITY FLORAL

A modern mix of fruits and flowers with added musk.
White Flowers..4 drops
Modern Fruits...1 drop
Musk...16 drops

ROSE CHYPRE

A classic chypre with enhanced floral heart.
Green Chypre...5 drops
Rose Accord..1 drops

WOODY FRUIT CHYPRE

A fruity chypre with an enhanced woody base.
Modern Fruits...1 drop
Woody...6 drops
Green Chypre...10 drops

EVENING WEAR

Presence and poise register with this group of perfumes; they are best suited to those occasions when your scent will be part of the atmosphere of the gathering. Delicate, composed and certainly suitable for the dinner party scene, intimate social occasion or gathering, these perfumes will provide that all-important final touch to your well-prepared appearance. Mysterious, radiant and seductive, these fragrances are often exotic and woody in nature.

ORIENTAL MUSK

A sultry oriental composition of woods and vanilla enhanced by musky notes.

Woody...11 drops
Musk..3 drops

GREEN FLORAL CHYPRE

A contemporary chypre enhanced by a touch of peach on a radiant floral heart.

Green Chypre..5 drops
Modern Floral..1 drop

WHITE CHYPRE

A sparkling floral bouquet harmonized by the freshness of galbanum on a warm oakmoss, ambery fond.

White Flowers......................................13 drops
Green Chypre..4 drops

MUSKY WOOD

A warm, woody, musky composition, with an exotic and seductive afterglow.

Musk..4 drops
Woody..3 drops

EVALUATING YOUR BLENDS

The best way to evaluate your fragrance blend is by means of vaporization. If you do not have a pump spray bottle, then place some of the blend on your wrist and smell it immediately to capture the top note. Periodically smell your wrist, keeping note of the changes in the scent and your reaction to it. You might want to ask a companion for his or her opinion as well. You can also try dipping a smelling strip into the blend. Is there a difference between the smell obtained from the smelling strip and the smell from the vaporized perfume? If so, which do you prefer? If it is the smelling strip version, then you will have to do some adjusting to your blend because it is the vaporized version that you will be wearing!

Perfume is most effective worn on warm skin: on the pulse points behind the ear, on the wrist and in the bend of the arm are good places. Clothes made of natural fibres are also good fragrance carriers. However, the perfume you use on your clothes can last, so if you like to change your fragrance often, it might be best not to spray the perfume directly on to your clothing unless you are planning to wash the item before wearing it again.

You will notice that some components appear in several of the complexes. This is due to the fact that there are many perfumes which rely on these particular notes. As fashions change, so does perfume. As mentioned earlier, even in classical Greece and Rome, perfumes were produced for various parts of the body. Today, our use of perfume is somewhat restricted to a wrist dab or a quick spray behind the ears or in front of the neck. This is just the starting point in the appreciation of perfumery; for now the emphasis is moving towards using perfume as a fashion tool to help you project the image you choose to the outside world. Perfume can be more than just a dab on the skin. It is an effective, discreet cloak which exists at a molecular level and can encompass the whole of one's being. This fine mist can be used on all parts of the body to provide a 'skin-wrap' which blends with one's particular 'aura' or personality.

Modern perfumes are composed using many compounds, natural and synthetic. The perfumes on the market today may be composed of up to 100 different materials, and it would be difficult to produce a variety of perfumes without a vast array of these materials. The fragrances in this kit have been selected in accordance with the aroma groups which are found in perfumes as discussed in Chapter Four.

The previous formulas provide an introduction to perfume blending, but, by varying the number of drops, you can make many more combinations according to your own personal preferences. Remember to write down your formulas, otherwise you may find it difficult to reproduce again. The seven complexes provided in this kit are also available from suppliers (*see Useful Addresses on page 111*).

MAKING BEAUTY PRODUCTS WITH YOUR KIT

To make cosmetics using the perfumes you have created with your kit, all you need is a basic unfragranced product. Your kit contains alcohol-based complexes. To some extent these will tend to dilute your cosmetics and in some cases the alcohol may change the composition. Should you wish to produce a large amount of a particular cosmetic, it would be better to have the original concentrate without the alcohol (*available from suppliers; see Useful Addresses page 111*). These concentrates will help you produce more stable cosmetics.

You can make a variety of cosmetics including skin creams, lotions, shampoos, bubble bath liquid and many others. It is possible to use essential oils as well if you particularly want to produce a personal range which can have both fragrant and therapeutic properties (*see Chapter Six*). The following are a few examples of cosmetics you can make.

CLEAR SKIN GEL

This is a good one to start with because it is very easy to make; unperfumed gels are very simple to work with and do not normally have an odour residue. Just a drop of perfume in a 50 g (2oz) pot produces a delicate, discreet fragrance. All you need to do is mix your perfume into the gel a drop at a time and leave for a few minutes to allow the perfume to blend. More perfume can be added if required, but it is wise not to overdo it. As the gel evaporates on the skin, it will leave a tiny, yet discernible glowing aroma. The gel acts like an Eau de Cologne; as the liquid or gel evaporates, it leaves a very thin film of perfume behind.

SKIN CREAM

Most of the unperfumed skin creams available on the market are made with a mineral oil base because it is more stable and has no residual smell.

Plant-based oils can be used, but they tend to have their own odour which may be a bit overpowering when blended with perfumes. To unperfumed cream add a few drops of your perfume at a time until there is a discreet aroma. This can be put into a glass jar and left to

stand. Remember that skin creams are generally supposed to be delicate in fragrance, due to the nature of the cream itself and the use to which it is put.

HAND LOTION

Using unperfumed hand or skin lotions, which are usually white, thin, runny liquids, add your drops of perfume and shake well. A bottle or jar can be used for this purpose. Allow the blend to settle, making sure that the mixture does not separate. Once the blend has set, there should be a delicate aroma rather than a strong perfume. When the cream is put on the skin, the fragrance is usually more pronounced.

DUSTING OR TALCUM POWDERS

A basic powder can be made by using unperfumed talc or by mixing 90 per cent unperfumed talc with chalk (which is calcium carbonate) and magnesium oxide. Talc, which is somewhat heavy, is soft and silky on the skin. Chalk is rather dry, but can be used in small amounts to lighten the powder. The magnesium oxide is very light and fine.

The dry powders should be carefully mixed in a large glass jar. To this jar, add your chosen perfume a drop at a time. Shake vigorously after each drop so that the drops or clumps of perfume disappear. Let it settle and check the smell. The perfume should just be noticeable without being overpowering. It you want it stronger, add a few more drops and repeat as above. It is a good idea to remove the lid after shaking, to allow any alcohol in the perfume to evaporate. Once all trace of the alcohol smell is gone, the powder can be packaged in a clean container or shaker for use.

SOLID PERFUME COLOGNE

Wax is used as the basis for these fragrant little tablets. They are a handy way of dispensing the perfume in a delicate manner. To prepare, melt a small amount of paraffin wax at a low temperature. This can be done by putting the wax in a Pyrex bowl or small saucepan and placing it in a larger saucepan partially filled with water over a low flame. Once melted, stir in your perfume, a drop at a time; add as many drops as you require to bring the mixture to the strength you desire. The mixture can then be poured into an old lipstick tube or small tin container. Allow to set.

SHAMPOO

Shampoos can vary greatly, and in the end it comes down to personal choice. It may be fun to experiment with making your own from this simple formula. Pour a small quantity of unperfumed shampoo – either clear or pearlized into a dish and add your perfume, a drop at a time. Stir and mix the liquid well, but do not whisk vigorously as this will cause unwanted air bubbles. Smell the blend to see if the perfume has been taken up and, if not, add more. If the shampoo seems to be thinning, add a pinch a salt. Once you have reached the desired fragrance, allow the shampoo to stand to make sure it is stable. It can then be poured into an empty plastic container, ready for use.

BUBBLE BATH

Bath accessories were originally designed to soften the water, but today they are mostly used to carry scent and make bathtime both fun and luxurious. Follow the recipe above for shampoos, using an unperfumed bubble bath liquid. And remember, fragrance is not just for the body. With the skilful use of the perfumes you have created, you can change the atmosphere in your boudoir, or any room for that matter, to project any mood you desire. Let your imagination guide your new-found skills.

Chapter Six

MOVING ON

*Where are they now, the days of aromatic warmth and
hot-scented remedies!*

GASTON BACHELARD

Now that
you have
a basic
understanding of
perfume blending
and have experi-
mented with creating
fragrances from the com-
plexes provided in your *Perfume
Kit*, you might like to recreate
some famous recipes from history or make your own adapta-
tions of these recipes. Many of
the recipes yield large quantities so you may wish to scale
down the recipes to make
smaller, more economical
amounts. An excellent way of
producing your own perfumes
is to use as a base a delicate
cologne which contains some
of the notes found in
the perfume you are
trying to create.
Essential oils of your
choice can then be
added, drop by drop,
to create your own
inspired formulas. Bear
in mind that pure essential
oils are very concentrated and
some essential oils are contra-
indicated for pregnant women
and those suffering certain
medical conditions. Some oils
can also cause skin irritation. It
is a good idea to do some pre-
liminary reading on aro-
matherapy beforehand (*see
pages 110–11 for a reading list
and suppliers of essential oils and
perfume bases*).

FORMULAS FROM HISTORY

Prior to the discovery of synthetic fragrances, there were many popular formulas for toilet waters and other fragrancing materials which have almost been forgotten. These formulas were made up of natural ingredients producing natural scents which are very much back in vogue these days and for that reason we have included some of those formulas from the past.

EAU DE COLOGNES

Basically, *Eau de Cologne* is a solution of approximately 3–5 per cent perfume oil in an alcohol/water base. Classic *Eau de Cologne* is a composition of citrus essential oils combined with other light and fresh oils. Only small amounts of fixatives are added; indeed, some formulas do not include them at all, which means that *Eau de Cologne* is not designed to last long. There are several different sorts of alcohol which can be used in the following recipes. Spirits of wine have long been considered to provide a very fine effect, however, rectified spirit from grain or potato can also be used. The choice depends on what type of fragrance you are trying to create.

Piesse felt that a better quality of *Eau de Cologne* would be produced by mixing the ingredients in a particular way, as indicated in his formula below. He also felt strongly that other ingredients should not be introduced, and that if it was necessary to produce a less expensive version, it was best to simply dilute the formula

with a weak spirit or rosewater. Note that all of the ingredients listed below (except the spirit) are essential oils.

PIESSE'S EAU DE COLOGNE

Rosemary Oil	28 ml (1 fl oz)
Neroli Petal	42 ml (1.5 fl oz)
Neroli Bigarade	14 ml (0.5 fl oz)
Orange Zest	70 ml (2.5 fl oz)
Citron Oil	70 ml (2.5 fl oz)
Bergamot	28 ml (1 fl oz)
Grape Spirit (60 per cent)	13.5 litres (3 gal)

First the citrine attars (citrus oils) should be added to the spirit and that mixture distilled. The rosemary and neroli oils should then be added to the distillate.

EAU DE COLOGNE II

Piesse also suggested a version of Eau de Cologne using a grain spirit (although he felt that the superior one was produced using grape spirit).

Lemon	113 ml (4 fl oz)
Bergamot	113 ml (4 fl oz)
Orange Peel Oil	113 ml (4 fl oz)
Petitgrain Oil	56 ml (2 fl oz)
Rosemary Oil	56 ml (2 fl oz)
Neroli Oil	14 ml (0.5 fl oz)
Grain Spirit (corn)	27 litres (6 gal)

Another version called *Eau de Cologne 18,* created by French perfumer I. T. Piver, consisted of about ten essential oils which were distilled in alcohol and filtered over fresh orange blossom.

OTHER CLASSIC EAU DE COLOGNE FORMULAS

The following recipe is a large one, in which the oils were measured in kilogrammes rather than litres. Essential oils of:

Bergamot..6.2 kg
Lemon...3.1 kg
Neroli..0.8 kg
Clove...1.6 kg
Lavender..1.2 kg
Rosemary...8 kg
Alcohol..100 litres

Another formula is based on a total percentage of 100 to allow for the preparation of small amounts if required.

Bergamot...33 per cent
Lemon..18 per cent
Sweet Orange...................................25 per cent
Petitgrain Bigarade...........................8 per cent
Lavender..6 per cent
Rosemary...5 per cent
Neroli...3 per cent
Clary Sage.......................................0.5 per cent
Benzoin Resinoid............................1.5 per cent

HUNGARY WATER

This was first produced in 1370 for Queen Elizabeth of Hungary.

Alcohol.....................................5 litres (1 gal)
Orange-flower Extract.............0.56 litres (1 pt)
Rose Essence...........................0.56 litres (1 pt)

Rosemary Oil..............................56 ml (2 fl oz)
Lemon Oil...................................28 ml (1 fl oz)
Melissa Oil..................................28 ml (1 fl oz)
Peppermint Oil..........................30 ml (1.1 fl oz)

Keep in a sealed container for a few weeks prior to use to allow oils to combine.

THE QUEEN'S PERFUME WATER

The following is another recipe prepared for Elizabeth of Hungary.

Rosewater, orange flower, sweet clover, myrtle flowers, costmary flowers, with a small amount of cinnamon and orange peel.

All the ingredients are placed together in a tightly sealed container for a month before use.

FLORAL BOUQUET

This cologne was traditionally used to perfume handkerchiefs.

Rose Extract................................1.7 litre (3 pt)
Violet Extract...............................1.7 litre. (3 pt)
Benzoin Tincture....................156 ml (5.5 fl oz)
Bergamot Oil...............................56 ml (2 fl oz)
Lemon Oil...................................28 ml (1 fl oz)
Orange Oil..................................28 ml (1 fl oz)

LAVENDER WATER

For this cologne, the plant materials and wine are distilled in a bain-marie. This recipe should yield about 1.7 litres (3 pt) of lavender water.

Lavender Flowers............................226 g (8 oz)
Rosemary Flowers.............................56 g (2 oz)
Wild Thyme.....................................56 g (2 oz)

Orange Flowers...............................85 g (3 oz)
Mint..113 g (4 oz)
Wine Spirit.................................3.4 litre (6 pt)

LOVE BOUQUET

This cologne was also used to perfume handkerchiefs.
Cassia Extract.............................1.1 litre (2 pt)
Jasmine Extract...........................1.1 litre (2 pt)
Rose Extract................................1.1 litre (2 pt)
Violet Extract..............................1.1 litre (2 pt)
Ambergris Extract....................0.56 litre (1 pt)
Musk Tincture...........................142 ml (5 fl oz)

BOUQUET DES CHASSEURS

Rose Essence...............................2.8 litre (5 pt)
Tonka Bean Tincture...................1.1 litre (2 pt)
Cassia Extract.........................0.56 litre (1 pt)
Neroli Extract.........................0.56 litre (1 pt)
Orange-flower Extract..............0.56 litre (1 pt)
Orris Root Tincture.................0.56 litre (1 pt)
Musk Tincture...........................284 ml (10 fl oz)
Lemon Oil.................................14 ml (0.5 fl oz)

FRANGIPANI

Bergamot.....................................1.18 ml (20 min)
Sandalwood.............................1.18 ml (20 min)
Neroli...0.88 ml (15 min)
Rose Geranium........................0.59 ml (10 min)
Rose..0.59 ml (10 min)
Verbena......................................0.30 ml (5 min)
Vetivert......................................0.30 ml (5 min)
Jasmine......................................142 ml (5 fl oz)
Tuberose....................................56 ml (2 fl oz)
Essence of Vanilla.......................28 ml (1 fl oz)
Benzoin Tincture.......................28 ml (1 fl oz)

TEA ROSE

Rose..5.3 ml (90 min)
Patchouli...................................88 ml (15 min)
Jasmine.......................................113 ml (4 fl oz)
Musk...56 ml (2 fl oz)
Tuberose....................................56 ml (2 fl oz)
Ambergris...................................28 ml (1 fl oz)
Orris...14 ml (0.5 fl oz)

All of the ingredients are blended in 0.56 litres (1 pt) of 80 per cent proof alcohol.

INCENSE

Used in ancient times as an offering to the gods, incense makes a wonderful room fragrancer.
Benzoin...226 g (8 oz)
Cloves...14 g (0.5 oz)
Cinnamon..two sticks
Calamus...one rhizome

All ingredients are pounded and sifted together, then mixed with tragacanth which has been dissolved in water, bringing the mixture to a paste-like consistency. It can then be formed into cone shapes or whatever is desired.

POMANDERS

The earliest known pomander comes from Greece and was of the bead type. Dioscorides offers a formula for scented lozenges known as rodides or rose pastilles in his Materia Medica. *The formula was very similar to later ones for pomander beads and is as follows:*
Fresh Roses *(before they become damp)* 40 drachmas
Spikenard..5 drachmas
Myrrh..6 drachmas

These are beaten fine and made up into little troches (small circular lozenges) which are then dried in the shade and stored in closely sealed jars. Sometimes honey and wine are included in the ingredients. The *troches* were worn around the neck to cover up the smell of sweat.

PREPARING YOUR OWN PERFUMES

Making perfumes using essential oils can be a fun and novel way of creating your own distinctive perfumes, either based on some great classical formula or simply one you have devised for yourself. You may encounter problems finding materials which are really essential when devising your own perfumes. If you want your perfumes to have some of the qualities which are present in many commerical perfumes today, you might need specific aroma chemicals (aldehydics), replacement animalic notes and certain balsams, which are not readily available in ordinary shops. You may also want to include other materials such as beeswax and resins as well; fortunately, all of these products are normally available from specialist supply houses (*see Useful Addresses page 111*).

ALCOHOL

Perhaps the most important ingredient in the art of perfume blending is alcohol. This is normally ethyl alcohol of perfumery grade, which has been denaturized to prevent it being taken internally, and is lacking the odour you would normally associate with ordinary alcohol. The type of alcohol you use may depend upon the other ingredients you will be using in your blend. Septimus Piesse felt that there was a marked distinction between British and French perfumes produced using the same recipes but with different alcohols. French perfumes using grape spirit had a much stronger bouquet due to the added fragrance that spirit contributes. Piesse felt that fragrances such as musk, ambergris, civet, violet, tuberose and jasmine should be made with grain spirit in order to retain their own aroma. However, he felt that blends such as *Eau de Cologne* worked better with grape spirit (brandy). As a guideline, if you want the fragrance of a particular material to come through strongly, use a grain alcohol which has little scent of its own.

OTHER INGREDIENTS AND TOOLS

You will also need some distilled or deionized water. Filter paper and funnels are useful when you want to clarify and clear any cloudiness which can appear. Toilet spirit, which is sold for beauty salons, is quite good as a dilutant but lacks the fineness of perfumery-quality spirit. There are some suitable *Eau de Cologne* formulations (weak colognes) available which can blend with the oils you are using; however, you need to experiment with them to be sure that they complement the particular fragrances you wish to use in your blend.

ESSENTIAL OILS

The use of natural products has become very popular and there are many choices available. The best natural ingredients for making perfumes are essential oils which have been extracted from a variety of plant parts; they can be mixed together to create the fragrances of your choice. Essential oils themselves are extremely concentrated and have therapeutic value as well as aromatic qualities. Some oils can irritate sensitive skin and some oils should be avoided during pregnancy. Refer to a good book on aromatherapy and essential oils if you are concerned about which oils might be safe to use. But they should only be used in small amounts diluted with a suitable carrier oil; for perfumery, jojoba or light coconut oil are suitable. Jojoba is an odourless oil with the consistency of a liquid wax and is a good conditioner for all skin types, which makes it a good choice to use as a perfume base. Light coconut oil is also non-oily and very spreadable, which makes it a good vehicle for essential oils.

For perfumes without alcohol, one or more of the floral essential oils (*see chart on page 96*) can be added to either light coconut or jojoba oil using a ratio of 4 drops of essential oil to 10 ml of carrier oil. Any of these florals on their own, or several blended together, can make a fragrant floral perfume.

For the more adventurous, some of the fruity or herby essential oils can be included. You may want to consider choosing one or more essential oils from each of the notes, just like perfumers do (*see notes chart page 43*). To make the fragrance last, at least one of the essential oils known for its fixative properties can also be added (*see chart page 96*). Fixatives generally fall into the base note category. However, one of the best natural fixatives is clary sage, as many of its components are exceptionally stable and persistent and its scent blends nicely with most essential oils. A tincture of clary sage can be prepared using alcohol and clary sage essential oil at a ratio of 1 part clary sage to 20 parts alcohol. This can be used in an *Eau de Cologne* or toilet water blend.

You can combine any number of essential oils, but remember to keep the ratio of essential oils to carrier oil as indicated above. The easiest way to assure this is to blend the desired essential oils together first until you achieve a desired scent, and then add the required number of drops of the blend to the carrier oil. Shake the mixture

---

SIMPLE HINTS FOR BLENDING PERFUME

1. *Have all your equipment clean and ready to use.*
2. *Write down what you are doing as you go along in your own still-room book.*
3. *Measure with care and remember that essential oils stick to glass.*
4. *Allow your creation to rest in a cool place for some days after preparation.*
5. *Never throw away a creation because you do not like it.*
6. *Remember that your perfume taste is personal to you.*
7. *Vaporization is the best way to smell your creation.*

ESSENTIAL OILS BY NOTE AND FRAGRANCE TYPE			
	TOP	MIDDLE	BASE
FLORALS		Chamomile, geranium, hyacinth, lavender	Jasmine, neroli, rose, tuberose, ylang ylang
FRUITY	Bergamot, grapefruit, lemon, lime, mandarin, orange, tangerine	Melissa	
HERB AND SPICE	Basil, coriander	Cypress, marjoram, rosemary	Cinnamon
FIXATIVES		Clary sage	Cedarwood, frankincense, patchouli, sandalwood

well. Perfumes using natural ingredients should be made in small amounts so that they can be used before they turn rancid. These perfumes should be kept in a cool, preferably dark place. Amber or coloured glass should also be used to help avoid unnecessary oxidation.

Once you have created your blend, you need to be able to bottle it. Perfume bottles are readily available from various suppliers (*see Useful Addresses, page 111*). This aspect is important if you are creating a perfume to give to a friend or even sell commercially. Bottles either come in open tops or with spray dispensers (known as pump sprays) with a screw-top bottle. The advantage of the pump sprays is that they are refillable from a larger container. Purse sprays are also available, which, as the name implies, conveniently fit into your purse or pocket. Purse sprays are usually made of anodized metal and hold quite small quantities of perfume – either 5 ml or 10 ml. They come in a tremendous range of shapes and colours, although basically they tend to resemble lipstick tubes. It is best to use a new container rather than one which had, or still has, a perfume in it. These sprays are difficult to clean and may smell of the previous contents.

SIMPLE FORMULAS TO MAKE

FLORAL WATERS

Floral waters are produced in three main ways: as a by-product of the distillation process using roses or other flowers, by using concentrates such as floral absolutes or by using terpeneless or carbon dioxide-distilled oils which have a better solubility in water. The problem with toilet waters is that they have to be sterilized by the addition of alcohol or a *biostat* (an ingredient which kills bacteria) to stop the solution from developing bacteria. Sometimes floral waters are alcoholic solutions, so the terms 'floral' and 'toilet' water tend to be interchangeable.

ROSEWATER

Make a mixture of distilled water or deionized water and alcohol mixed at the ratio of 90 per cent water and 10 per cent alcohol. (Deionized water is available at garages and at some pharmacies.) Mix and stir, then add rose absolute a drop at a time. Stir constantly until the drops are fully incorporated. More oil can be added until the water becomes more fragrant, usually at a rate of 3 drops per 100 ml. Leave for some hours and repeat the stirring until all the rose absolute is incorporated. If there is any left floating on the top, remove by blotting with a tissue. Use immediately or bottle. If you intend to store it, add a bacterial inhibitor (these are available from supply houses). You can achieve a similar effect by adding a few drops of non-fragrant detergent to your mixture.

FRAGRANT FLORAL WATER

To your basic water and alcohol mixture, add drops of terpeneless petitgrain oil, rose absolute and tuberose absolute (up to 3 per cent of the total mixture). Do not exceed the safe dilution ratio for essential oils (*for guidelines regarding essential oils, see page 95*). You can also add terpeneless orange oil if you wish. Mix the liquids by stirring until all the oils have dispersed. Leave to settle before use.

LAVENDER WATER

To your basic water and alcohol mixture, add drops of terpeneless or carbon dioxide-extracted lavender oil (between 3 and 5 per cent of the total mixture). Stir until all is incorporated. Use as a skin freshener immediately or bottle and store bearing in mind that you need to add a bacterial inhibitor.

LISBON WATER

Though true *Lisbon Water* uses 100 per cent proof alcohol, it is possible to produce a passable version of this water using a weaker alcohol.

To your basic mixture add drops of terpeneless orange oil and lemon oil at the ratio of 2 parts orange oil to 1 part lemon oil. Mix until well blended, then add a drop or two of rose absolute. Stir and mix this blend until fully incorporated. Either use immediately or bottle in a sterilized container and add a bacterial inhibitor if you plan to store it for any length of time.

EAU DE COLOGNES AND TOILET WATERS

Eau de Colognes and toilet waters are normally made of ethyl alcohol with small amounts of essential oils added. Many traditional recipes rely on the fresh notes of the citrus oils. Listed below are a few examples. Please note that 1 ml equals 20 drops of essential oil. It is possible to obtain a pipette which will extract 1 ml amounts of your chosen essential oil very easily from a bottle. For the recipes below, first blend the essential oils together then add the alcohol before bottling.

EAU DE LISBON

Lemon	2 ml
Orange	4 ml
Rose	4 drops
Alcohol	*up to* 100 ml

HUNGARY WATER

Lemon	1 ml
Melissa	1 ml
Peppermint	1 ml
Rosemary	1 ml
Alcohol	*up to* 100 ml

ALCOHOL-BASED PERFUMES

These perfumes are more complex fragrances requiring several more essential oils than the previous waters, but they also rely on alcohol as a base. Blend the essential oils first, then add the alcohol to the oils.

THE ALHAMBRA PERFUME

Tuberose Absolute	4 ml
Geranium	2 ml
Neroli	2 ml
Orange	2 ml
Rose Absolute	2 ml
Hyacinth Absolute	1 ml
Alcohol	*up to* 100 ml

ROYAL BOUQUET

Jasmine Absolute	2 ml
Tuberose Absolute	2 ml
Lavender	1 ml
Cassia	1 ml
Rose	3 ml
Rose Absolute	2 ml
Alcohol	*up to* 100 ml

JOCKEY CLUB BOUQUET

Rose	4 ml
Tuberose Absolute	4 ml
Cassia	2 ml
Jasmine Absolute	4 ml
Bergamot FCF	1 ml
Alcohol	*up to* 100 ml

MILLEFLEURS

Tuberose	2 ml
Jasmine Absolute	2 ml
Violet Absolute	1 ml
Rose	3 ml
Cassia	1 ml
Neroli	2 ml
Vanilla Absolute	1 ml
Cedarwood	1 ml
Geranium	1 ml
Alcohol	*up to* 100 ml

Bouquet of all Nations

Morocco: Rose Absolute............................2 ml
India: Jasmine Absolute..............................2 ml
Indonesia: Patchouli....................................2 ml
France: Tuberose Absolute.........................1 ml
Brazil: Vanilla Absolute..............................1 ml
USA: Orange...1 ml
Italy: Lemon..1 ml
Spain: Lavender..1 ml
Bulgaria: Rose Otto....................................2 ml
Sri Lanka: Citronella..................................1 ml
Madagascar: Ylang Ylang..........................3 ml
Alcohol...*up to* 100 ml

Bouquet de la Reine

Rose Otto...4 ml
Tuberose..3 ml
Neroli...3 ml
Bergamot FCF...3 ml
Rose Absolute...3 ml
Alcohol...*up to* 100 ml

Rondeletia Perfume

There are several versions of this perfume, but its base is a mixture of lavender and clove. The blending of the perfume can be tricky due to the variable nature of the oils used; these particular essential oils tend to vary in strength and intensity from region to region and from season to season. You will have to experiment a little with this particular recipe until you have a a fragrance that pleases you.

Lavender..8 ml
Clove..4 ml
Bergamot FCF...4 ml
Rose Otto...2 ml

Modern Fragrances

Listed here are a few newer fragrances which take into consideration the wide range of essential oils and absolutes which are available today.

Tropical Flowers

Jasmine Absolute...3 ml
Ylang Ylang...2 ml
Rose Absolute...4 ml
Patchouli..2 ml
Vanilla Absolute..2 ml
Vetivert..2 ml
Alcohol...*up to* 100 ml

Indian Magic

Sandalwood..6 ml
Jasmine Absolute...8 ml
Tuberose Absolute.......................................3 ml
Alcohol...*up to* 100 ml

Bouquet des Fleurs

Bergamot FCF...6 ml
Lemon..4 ml
Rose Otto...4 ml
Tuberose Absolute.......................................2 ml
Alcohol...*up to* 100 ml

Bouquet des Orientales

Rose...4 ml
Cedarwood Atlas..2 ml
Patchouli..2 ml
Sandalwood..2 ml
Verbena..2 ml
Vetivert..1 ml
Alcohol...*up to* 100 ml

MASCULINE PERFUMES AND AFTER-SHAVES

The following colognes and after-shaves are some suggested formulas for the special man in your life. Some rely on the traditional citrusy, woody notes popular in men's perfumes. Again, mix the essential oils first, then add the alcohol.

SPLASH-ON COLOGNE

Sandalwood	4 ml
Bergamot FCF	2 ml
Alcohol	*up to* 100 ml

WOODY COLOGNE AFTER-SHAVE

Cedarwood Atlas	3 ml
Cypress	1 ml
Lemon	1 ml
Rosemary	1 ml
Sandalwood	2 ml
Alcohol	*up to* 100 ml

FRESH COLOGNE AFTER-SHAVE

Bergamot FCF	3 ml
Orange	1 ml
Rosemary	1 ml
Alcohol	*up to* 100 ml

A MAN'S FRAGRANCE

Sandalwood	3 ml
Patchouli	2 ml
Cedarwood	4 ml
Bergamot FCF	2 ml
Vetivert	1 ml
Frankincense	1 ml
Alcohol	*up to* 100 ml

SPICY BOUQUET

Hyacinth Absolute	2 ml
Tuberose Absolute	1 ml
Linden Blossom Absolute	2 ml
Rose Absolute	3 ml
Cassia	1 ml
Sandalwood	2 ml
Patchouli	2 ml
Alcohol	*up to* 100 ml

PERFUMED SKIN CREAM

The following is an easy-to-make formula for skin cream from natural ingredients.

Grated Beeswax	2 tablespoons
Almond Oil	20 ml (4 teaspoons)
Distilled or Spring Water	20 ml (4 teaspoons)
Essential Oil	8 drops

(You can use more than one essential oil as long as the total amount used does not exceed 8 drops.) You can also use any other perfume blend of your choice.

To prepare the skin cream, first combine the shredded beeswax and almond oil in a Pyrex dish or small saucepan and place it in a pan of hot water over a low heat. Stir the mixture carefully until the beeswax has melted completely. Remove from the heat and add the water and essential oils (or other perfume blend) while whisking the melted wax rapidly to emulsify the mixture. Pour immediately into a sterilized jar with a tight-fitting lid. This cream can be kept for about a month without refrigeration. Beeswax can be obtained from health shops and from good craft shops (*see page 111 for suppliers and useful addresses*).

OTHER USES FOR PERFUMES

Once you have created your own exclusive perfume, you can use it as your own signature for many things. And if you have devised several perfumes, you might want to use them for special effects throughout your home.

Perfumes can be used to add fragrance to an almost endless number of things, with the only limit being your imagination. In fact, there are many things in our environment that are purposefully fragranced which we may not even be consciously aware of. The leathery fragrance of a new car interior, for example, has actually been bottled, and used-car dealers spray it in cars to lull the would-be purchaser into a false sense of security about a car that might not be in the best mechanical condition.

There are a number of ways you can use fragrance to enhance your environment or add a pleasant note to a social situation. The following are just a few suggestions.

ROOM FRAGRANCER

Rooms do not take kindly to perfumes. When Cleopatra used rose perfume to dramatic effect in her love life, she made sure that the door was closed before she sprayed her favourite perfume on her bed. The secret is to use them for effect just a few moments before you need to make the fragrant statement. Do not open the doors and windows after spraying, otherwise the precious aroma will simply blow away.

Instead of using your spray perfume like an air-freshener, dab or spray perfume on cards which can be placed at strategic positions throughout the room. This allows the perfume to slowly release its fragrance.

Another possibility is to have dangling mobiles which can be sprayed from time to time. Alternatively a few well-placed handkerchiefs with a generous helping of perfume will be effective. Warm materials, like fabric, take up perfumes better than paper or card. Later on you can spray discreetly, if you find a positive response to the fragrance.

VAPORIZERS

A perfume blend made from natural essential oils is best to use in one of the many vaporizers available. These vaporizers usually consist of a small dish designed to hold water to which 4 to 6 drops of your essential oil fragrance may be added. The water and oil mixture is heated by a nightlight candle, although some models incorporate an electric bulb. The fragrance is dispersed as the water and oil mixture evaporates.

If you do not have a vaporizer, 4 to 6 drops of your perfume blend can be added to a piece of dry cotton wool and placed directly on a radiator. Drops from an essential oil blend can also be added to humidifier containers which disperse moisture in a room. Adding fragrance to a bowl of boiling water will also work, but the scent will not last quite as long as any of the other methods mentioned.

SCENTED CANDLES

There are candle kits available which provide the moulds, essential ingredients and instructions for making candles. After the beeswax and paraffin have been melted, the desired number of drops of your perfume blend can be added just prior to pouring the mixture into the mould.

If you do not want to make your candles from scratch, you can scent existing candles. Light the wick, allowing some of the wax to melt before carefully placing 3 or 4 drops of your perfume blend into the melted wax. After just a few minutes a delicate fragrance will fill the area.

POT-POURRIS AND SACHETS

A sachet powder can be prepared using almost any fragrant dried flower, herb or spice. As some spices are already dried when bought, it is a simple matter to adapt them for your use. Take any available aromatic plant material – leaves, flowers, herbs and so on. This is best done on a warm summer day when the plant petals or leaves are dry. Collect them in a dish and place them in a warm room on muslin where the air can circulate. When completely dry, place into a large closed jar or other container. They should be crispy dry and have no moisture at all so that they can be blended into a fine powder. Into this powder you then need to add your choice of resin, such as powdered frankincense, myrrh or other material. Add to this a certain amount of orris powder into which you have previously added your perfume. Allow the powder to dry out completely before you add to the main mixture. Stir and leave for a few days. Now make some little cotton sachets which you can fill with the powder and then tie with a ribbon. Dried plant petals or leaves tend to have a tea-like smell but you can change this by adding spices such as cloves and cinnamon.

A typical recipe from a century ago reads as follows: mix together ground cedarwood, sandalwood and rosewood. Mix and sift and then add a few drops of rose otto. It is possible to reproduce this formula using finely sifted sawdust from your local timber yard. This already may have an interesting odour. Then you can add a few drops of sandalwood, howood and cedarwood and some rose otto. Shake and allow to become a fragrant powder ready to sew into little silk pouches.

Once you have created one type, you can move on to other combinations. Mutio Frangipani, a noted Italian alchemist, created a special sachet powder which still bears his name. The recipe is as follows: three parts orris root powder to one part sandalwood powder and one part vetivert powder. Mix together and then add several drops of sandalwood, rose otto and neroli. Blend the mixture. The original powder contained a small amount of musk, but as this is no longer used, you can add a musk seed or two which smells rather like musk.

Sachets can be made out of any fragrant ingredient – all you have to remember is to dry the material carefully, blend it into a powder and then add your aromatics. Lavender flowers, rose leaves, ground cloves and ground cinnamon are some suggestions for materials that can be used.

Several drops of your favourite perfume blend can be added to refresh a pot-pourri or sachet. A bowl filled with pot-pourri is a nice touch for a bathroom, bedroom or anywhere in the house where you want to add a bit of decoration and scent. Sachets are ideal for your drawers or closet, imparting a gentle scent to your lingerie, blouses and other articles of clothing.

LETTERS AND NOTES

Special letters to friends or lovers can be perfumed, though you need to check periodically to make sure that the odour survives well. The best way to perfume paper is to keep all your stationery – envelopes and writing paper – in a box into which you have placed a piece of paper, kitchen paper or a tissue with some of your perfume. The aroma molecules will slowly spread through the contents and give a delicate fragrance to the paper. If you have created a perfume which does not last very long, add an extra drop before you seal the letter, to provide a fragrant finishing touch.

Of course if you are in the cosmetic, perfume or aromatherapy business, it may be advantageous to announce yourself before your letters are opened. In fact, it is now possible to have perfumed inks, and to give perfumed calling cards or letters. It may not work with the bank manager if you want to borrow money, but it can charm most other people.

SCENTED WADDING OR FABRICS

Pieces of scented wadding or fabric can be used to provide a small amount of scent to be released in special places. Scented wadding made out of cotton wool can be covered with a piece of coloured material to add to a linen drawer. Similarly, they can be used in pin cushions, workboxes and so on. These can even be used in a car where you may want to get away from the more masculine smells of leather. Indeed most car leather cleaning sprays tend to have strong leather notes.

These are just a few ideas. With a little imagination you will be able to use your favourite perfumes in many new ways.

APPENDIX
FEMININE FRAGRANCES

FLORAL PERFUMES

A by Annabella / Annabella-Denis
Alfred Sung / Riviera Concepts
Amarige / Givenchy
Annabella / Denis
Anne Klein / Parlux
April Violets / Yardley
Aramis 900 / Aramis
Aria / Missoni
Ariane / Avon
Arnold Scasi / Revlon
Asja / Fendi
Azzaro 9 / Azzaro
Barynia / Helena Rubinstein
Beautiful / Estée Lauder
Bellodgia / Caron
Bill Blass / Revlon
Blue Grass / Elizabeth Arden
Brise de Printemps / Mibelle
Bulgari / Bulgari
Cabotine / Grès
Calyx / Prescriptives
Capricci / Nina Ricci
Cardin / Pierre Cardin
Carolina Herrera / Herrera-Puig
Cassini Ladies / Cassini
Chanel No. 22 / Chanel
Charivari / Charles of the Ritz
Chassres / Pola
Chess d'Or / Chess
Chloé Narcisse / Parfums Chloé
Cornubia / Penhaligon's
Courrèges in Blue / Courrèges
Création / Lapidus
Dans la Nuit / Worth
Davana / Berger
Di Borghese / Borghese
Dilys / Laura Ashley
Diorama / Christian Dior
Eau de Fleurs / Nina Ricci
Eau de Gucci / Gucci
Elizabethan Rose / Penhaligon's
English Flowers / Taylor of London
English Rose / Yardley
English Spring / Yardley
Envol / Lapidus
Escada / Escada-Margretha Ley
Eternity / Calvin Klein
Facets / Avon
Fantasque / Feraud-Avon
Ferrose / Ferro
Fête des Roses / Caron
Fleur / Lenthéric
Fleurs à Fleurs / Diparco
Fleurs de Rocaille / Caron
Fleur d'Interdit / Givenchy
Flora Danica / Swank-Worldwide
Freesia / Yardley
Francesca / Mibelle
Gardenia / Chanel
Gardenia Passion / Goutal
Genny / Hanorah
Germaine / Monteil
Gianfranco Ferré / Ferré-de Silva
Giorgio Beverly Hills / Giorgio Beverly Hills
Goce-Goce / Morris
Gran Palais / Raison Pure
Gucci No. 1 / Gucci
Gucci No. 3 / Gucci
Guirlandes / Carven
Habanita / Molinard
Hanorable / Hanorah
Happy Diamonds / Chopard
Heure Exquise / Goutal
Heure Intime / Vigny
Histoire d'Amour / Aubusson
Honeysuckle / Avon
Impromptu / Avon
Insolent / Jourdan
Intrigue / Carven
JA / Jun Ashida
Jardins de Bagatelle / Guerlain
J'Arrive / Mibelle
Jasmin / de Varens
Jasmin de Corse / Coty
JD / Jesse Daniel
Jour / Feraud-Avon
Joy / Jean Patou
La Belle Vie / Sacha Distel
Laguna / Salvador Dali
L'Aimant Eternelle / Coty
La Rose Jacqueminot / Coty
Laura Ashley No. 1 / Laura Ashley
Lavender / Floris
Lavender / Woods of Windsor
L'Eau du Soir / Sisley
Le Baiser du Faune / Molinard
Le Dandy / d'Orsay
Le Nouveau Gardenia / Coty
Le Vertige / Coty
Lily / Floris
Lily of the Valley / Floris
Lily of the Valley / Woods of Windsor
Lily of the Valley / Yardley
L'Or / Coty
Lumière / Rochas
Madrigal / Molinard
Magnolia / Rocher
Mai / Shiseido
Maissa / Vermeil
Marbert Woman / Marbert
Maxims de Paris / Pierre Cardin
Métamorphose / Laporte-L'Artisan Parfumeur
Météor / Coty
Millefleurs / Crabtree & Evelyn
Mon Classique / Morabito
Moschus Cool Love / Aok-Nerval
Moschus Magic Love / Aok-Nerval
Moschus Wild Love / Aok-Nerval
Muse / Coty
Naj Oleari / Euroitalia
Niki de Saint Phalle / PPI
Nina / Nina Ricci
Nino Cerruti pour Femme / Cerruti
Nuit de Longchamp / Lubin
Number One / Betrix
Odyssey / Avon
Ombre d'Or / J. C. Brosseau
Only / Iglesias-Myrurgia
Open / Roger et Gallet
Ophelia / Avon
Orfea / de Varens
Or Noir / Morabito
Parfum de Femme / Goutal
Parfum d'Elle / Montana
Paris / Coty
Paris / Yves St Laurent
Passion / Goutal
Perry Ellis / Ellis-Stern
Petunia / Yardley
Poésie / 4711
Prêt à Porter / Perfumers Workshop
Princess d'Albret / d'Albret
Ptisenbon / Tartine et Chocolat-Givenchy
Red Door / Elizabeth Arden
Reflections / Avon
Rivage / Shiseido
Roméo / Romeo Gigli
Rose / Czech & Speake
Rose Absolue / Goutal
Rose Cardin / Pierre Cardin
Rose de Bulgari / Bulgari
Rose de Noël / Caron
Rose de Rouge / Gemey
Rose Roses / Yardley
Rosewater / Crabtree & Evelyn
Samba / Perfumers Workshop
Saphir / de Varens
Selection / Lancaster
Sfera / Faux
Sinan / Sinan
Sithonia / Mibelle
Society / Burberry
Sophia / Coty
Sourire / Shiseido
Spectacular / Joan Collins
Stephanie / Bourjois
Sun, Moon, Stars / Lagerfeld
Sung / Sung
Suzuro / Shiseido
Symboise / Stendhal

Tatiana/von Furstenberg
Tea Rose/Perfumers Workshop
Teatro alla Scala/Krizia
Tendre Poison/Dior
Torrrente/Torrente
Traditional/Woods of Windsor
Tristano Onofri/Babor
Trussardi Action/Trussardi
Tubéreuse/Goutal
Tuberose/Chess
Ungaro d'Ungaro/Ungaro
Valentino/Valentino-Stern
Vicky Tiel/Tiel
Victorian Posy/Penhaligon's
Vie Privée/Rocher
Volcan d'Amour/Diana von
Furstenberg
Votre/Jourdan
VSP/Jovan
White Diamonds/Elizabeth Taylor
White Satin/Yardley
White Shoulders/Evyan
Wild Rose/Woods of Windsor
Wind Song/Matchabelli
Windsor Blossom/Woods of
Windsor
Wings/Giorgio Beverly Hills
Woman III/Jil Sander
Xmas Bells/Molinard
XS pour Elle/Paco Rabanne
Yolia/Cantilene
You're the Fire/Yardley
Zen/Shiseido
Zinnia/Floris
1000/Jean Patou

FLORAL-SWEET PERFUMES

Adolfo/Denney
Afghane/d'Estrées
Babe/Fabergé
Bijan Perfume/Bijan
Blasé/Max Factor
Blue Moon/Mibelle
Boucheron/Boucheron
Cabriole/Elizabeth Arden
Candid/Avon
Capucci/Capucci
Cerissa/Revlon
Cerruti 1881 Femme/Cerruti

Champagne/Monteil
Charles of the Ritz/Charles of the
Ritz
Chloé/Parfums Chloé
Enjoli/Charles of the Ritz
Fracas/Piguet
Garconne/Illuster
Ginseng/Jovan
Halston Night/Halston
Island Gardenia/Jovan
Jontue/Revlon
Jungle Gardenia/Tuvache
Kalispera/Desses
Lady 80/Kanebo
Lalique/Lalique
La Madrague/B. Bardot
L'Heure Bleue/Guerlain
Liberty/Yardley
L'Origan/Coty
Marilyn Monroe/East West-
Colorkit
Michelle/Balenciaga
Miss Worth/Worth
Moon Drops/Revlon
Oscar de la Renta/Sanofi
Pavlova/Pyot
Première/Castelbajac-4711
Ruffles/Oscar de la Renta
Soir de Paris/Bourjois
Stephanotis/Floris
Suggestion/Montana
Sweet Honesty/Avon
Unspoken/Avon
Vanderbilt/Gloria
Venezia/Laura Biagiotti
Via Lanvin/Lanvin
Yendi/Capucci

FLORAL-FRUITY PERFUMES

Allora/Marbert
Anaïs Anaïs/Cacharel
Anthracite/Jourdan
Armani/Giorgio Armani
Bambou/Weil
Baruffa/Atkinsons
Cassini/Cassini
Chant d'Arômes/Guerlain
Charade/Max Factor
Chiara Boni/Boni

Clair de Jour/Lanvin
Clin d'Oeil/Bourjois
Dalissime/Salvador Dali
Duende/Duende
Eau d'Azzaro/Azzaro
Eau de Gucci/Gucci
Eau Fraiche/Léonard
Eden/Cacharel
Envol/Lapidus
Estivalia/Puig
Exclamation/Coty
Impress/Kanebo
Isa/de Varens
Jean Louis Sherrer/Jean Louis
Sherrer
Jitrois/Jitrois
Kenzo de Kenzo/Kenzo
Lauren/Ralph Lauren
Le Jardin/Max Factor
Le Sport/Coty
Mary Quant/Max Factor
Miss Arpels/Van Cleef & Arpels
Miss Habanita/Molinard
Ô de Lancôme/Lancôme
Panache-Evening/Lenthéric
Peach Blossom/Woods of Windsor
Quartz/Molyneux-Sanofi
Red Jeans/Versace
Régine's/Interperfume
Rumba/Balenciaga
Scoundrel/Revlon-Collins
Sweet Courrèges/Courrèges
Talisman/Balenciaga
Tiffany/Tiffany
Trésor/Lancôme
Tribu/Benetton
Turbulences/Révillon
Ungaro/Ungaro
Un Jour/Jourdan
Vison/Beaulieu
Vivid/Liz Claiborne
Volupté/Oscar de la Renta
White Flowers/Astor
Wild Orchid/Woods of Windsor
273/Hayman
360°/Perry Ellis

FLORAL-FRESH PERFUMES

Balahé/Léonard

Balestra/Balestra
Byblos/Hanorah
Casaque/d'Albret
C'est Moi/Aigner
Charlie/Revlon
Cristalle/Chanel
Diorella/Christian Dior
Diorissimo/Dior
Elle/Lenthéric
Emprise/Avon
Fidji/Laroche
First/Van Cleef & Arpels
Flamme/Bourjois
Foxfire/Avon
Grain de Sable/Verfaillie
Hanae Mori/Shiseido
Inspire/4711
It/Lenthéric
Jeu de Fleurs/Mibelle
Jordache Woman/Jordache
Judith/Müller
K de Krizia/Krizia
Khadine/Yardley
Kiry/Marbert
Lady A/Alcina
Laughter/Tuvache
Laura Biagiotti/Biagiotti-Betrix
L. de Lubin/Lubin
Le Muguet du Bonheur/Caron
Les Fleurs/Houbigant
Lieu du Blanc/Pola
Madame de Carven/Carven
Metal/Paco Rabanne
Muguet des Bois/Coty
My Melody/4711
Norell/Revlon
Number Two/Betrix
Parfum d'Été/Kenzo
Rafale/Molinard
Ritz/Charles of the Ritz
Seringa/Floris
White Lilac/Chess

GREEN PERFUMES

Ainsi/Atkinsons
Alada/Myrurgia
Alliage/Estée Lauder
Après l'Ondée/Guerlain
Babore/Babor
Basile/Sirpea

Bluebell / Penhaligon's
Byblos / Byblos
Calvin Klein / Calvin Klein
Chanel No. 19 / Chanel
Clinique / Clinique
Dune / Christian Dior
Eau de Fraicheur / Weil
Escape / Calvin Klein
Espiegle / Atkinsons
Graffiti / Capucci
Inouï / Shiseido
Ivoire / Balmain
Janine D / 4711
Jil Sander / Jil Sander
Joker / Nerval
Lancetti / Lancetti-Denis
L'Eau D'Issey / Issey Miyake
Loewe / Loewe
Mademoiselle Ricci / Nina Ricci
Mimosa / Czech & Speake
Murasaki / Shiseido
New West for Her / Aramis
Omar Sharif pour Femme / Prestige
Pheromone / Miglin
Private Collection / Estée Lauder
Quant by Quant / Quant
Ravissa / Maurer & Wirtz
Sabatini / G. Sabatini
Safari / Ralph Lauren
Sex Appeal for Women / Jovan
Shocking You / Schiaparelli
Silences / Jacomo
Silverline / Gainsborough
Sport Scent for Women / Jovan
Style / Avon
Sunflowers / Elizabeth Arden
Tamore / Nerval
Tramp / Lenthéric
Vent Vert / Pierre Balmain
Veronese / de Varens
Vivara / Pucci
Weil de Weil / Weil
Wrappings / Clinique
Y / Yves St Laurent

ALDEHYDIC PERFUMES

Adeline / Pola
Antilope / Weil
Arpège / Lanvin
Aviance / Matchabelli

Babillage / Kanebo
Baghari / Piguet
Berlin / Joop
Bizarre / Atkinsons
Bois des Iles / Chanel
Calandre / Paco Rabanne
Calèche / Hermès
Chamade / Guerlain
Chanel No. 5 / Chanel
Chantage / Lancaster
Charisma / Avon
Chicane / Jacomo
Chunga / Weil
Cléa / Rocher
Climat / Lancôme
Coeur-Joie / Nina Ricci
Complice / Coty
Coryssima / Coaryse Salomé
Detchema / Revillon
Donna R / di Camerino
Ecusson / d'Albret
Edwardian Bouquet / Floris
Elegance / Avon
Embrujo de Sevilla / Myrurgia
Evasion / Bourjois
Exploit / Atkinsons
Fashion / Léonard
Fleurs de Fleurs / Nina Ricci
Florissa / Floris
Forget-Me-Not / Woods of Windsor
French Style / Betrix
Gauloise / Molyneux
Gold / Lenthéric
Halston / Halston
Havoc / Quant
Hinotori / Kanebo
Hope / Denney
Imprévu / Coty
Infini / Caron
Jardanel / Desprez
Je Reviens / Worth
Joya / Myrurgia
Kyoto / Kanebo
Lady / Jovan
Le Dix / Balenciaga
Le Jardin D'Amour / Max Factor
l'Interdit / Givenchy
Liu / Guerlain
Lutèce / Houbigant
Madame Jovan / Jovan

Madame Rochas / Rochas
Maxi / Max Factor
Mémoire Chérie / Elizabeth Arden
Miss de Rauch / de Rauch
Misty Tea Rose / Jovan
Moodwind / Avon
More / Shiseido
Mystique / Lenthéric
Nahema / Guerlain
Nishiki / Shiseido
Nocturnes / Caron
Nombre Noir / Shiseido
Nonchalance / Maurer & Wirtz
Nuance / Coty
Nude / Bill Blass
Oleg Cassino for Women / Jovan
Ombre Rose / Brossard
Opéra / Coryse Salomé
Orgia / Myrurgia
Panache / Lenthéric
Prophecy / Matchabelli
Révillon 4 / Révillon
Risque / Juvena
Rive Gauche / Yves St Laurent
Robe d'un Soir / Carven
Sa So / Shiseido
Soft Musk / Avon
Super Fragrance / Aigner
Surreal / Avon
Tamango / Léonard
Tasha / Avon
Tendance / Marbert
Tendresse / Atkinsons
Tigress / Fabergé
Topaze / Avon
Tosca / 4711
Touche / Jovan
Tweed / Lenthéric
Uninhibited / Cher
Vivre / Molyneux
White Linen / Estée Lauder
Yesterday / Marbert
Zibeline / Weil

CHYPRE PERFUMES

Amérique / Courrèges
Andron / Jovan
Animale / S. de Lyon
Aphrodisia / Fabergé
Apogée / Les Senteurs

Aquarius / Max Factor
Azurée / Estée Lauder
Azzaro / Azzaro
Bandit / Piguet
Bat Sheba / Müller
Black Velvet / Yardley
Cabochard / Grès
Cachet / Matchabelli
Care No. 2 / Astor
Casanova / Casanova
Catherine Deneuve / Phoenix
Chimère / Matchabelli
Chique Silver / Yardley
Choc / Pierre Cardin
Champagne / Yves St Laurent
Chypre / Coty
Cialenga / Balenciaga
Ciao / Houbigant
Coriandre / Couturier
Courant / Helena Rubinstein
Crêpe de Chine / Millot
Daydreams / Maybelline
Deci-Delà / Nina Ricci
Demi-Jour / Houbigant
De Soir / Betrix
Diamella / Rocher
Dune / Christian Dior
Durer / Durer
Eleven / Atkinsons
Empreinte / Courrèges
Epris / Max Factor
F. de Ferragamo / Fendi
Farouche / Nina Ricci
Femme / Rochas
Fête / Molyneux
Flair / Yardley
Forever Krystle / Charles of the Ritz
Geminesse / Max Factor
Gianni Versace / Versace
Givenchy III / Givenchy
Glamour / Bourjois
Halston Couture / Halston
Hascish / Veejaga
Intimate / Revlon
Intuition / Max Factor
Italian Style / Betrix
Jacaranda / 4711
Jolie Madame / Balmain
Knowing / Estée Lauder
Koto / Shiseido

Lace/Yardley
La Nuit/Paco Rabanne
L'Arte/Gucci
Le Parfum/Sonia Rykiel
Le Temps d'Aimer/Alain Delon
Ma Griffe/Carven
Marc Aurele/Mibelle
Masumi/Coty
Mila Schon/Schon
Mink & Pearls/Jovan
Miss Balmain/Pierre Balmain
Miss Dior/Christian Dior
Miss Factor/Max Factor
Missoni/Missoni
Misty Wind/Mibelle
Mitsouko/Guerlain
Moments/Colonia-Mulhens
Monoa/Mibelle
Mon Parfum/Paloma Picasso
Montana/Montana
Mystère/Rochas
Nipon/Nipon
Nitchevo/Juena
Norell II/Revlon
Nueva Maja/Myrurgia
Occur/Avon
Or/Torrente
Paloma Picasso/Paloma Picasso
Paradoxe/Pierre Cardin
Parfum Rare/Jacomo
Partage/Fabergé
Parure/Guerlain
Pink Lace/Yardley
Pure Silk/Yardley
Quadrille/Balenciaga
Red/Beene
Reverie/Tuvaché
Sculptura/Jovan
Secret de Venus/Weil
Senchal/Charles of the Ritz
Senso/Camarino

Septième Sens/Sonia Rykiel
Shalom/Müller
Shéhérazade/Desprez
Sikkim/Lancôme
Sous le Vent/Guerlain
Style/Lenthéric
Timeless/Avon
Trussardi/Trussardi
V E/Charles of the Ritz
Visa/Piguet
Woman Two/Jil Sander

ORIENTAL PERFUMES

Amouage/Amouage
Amun/4711
Angel/Thierry Mugler
Bal à Versailles/Desprez
Bouvardia/Floris
Breathless/Avon
Byzance/Rochas
Cananga/Berger
Carnet du Bal/Révillon
Casmir/Chopard
C'est la Vie/Lacroix
Chantilly/Houbigant
Ciara/Revlon
Ciel/de Varens
Cinnabar/Estée Lauder
Cinnamon & Orange/Woods of
Windsor
Coco/Chanel
Colors/Benetton
De Jour/Betrix
Desire/Max Factor
Dioressence/Christian Dior
Eau de Caron/Caron
Emeraude/Coty
Enigma/de Markoff
Expression/Fath
Fémininité du Bois/Shiseido
Frankincense & Myrrh/Jovan

Grain de Folie/Verfaillie
Haute Parfumerie/Jean-Paul
Gaultier
Herbessence/Helena Rubinstein
Interlude/Denney
Ispahan/Rocher
J'ai Osé/Guy Laroche
Jaipur/Boucheron
Jicky/Guerlain
Joop!/Joop
Kashaya/Kenzo
Keora/Couturier
Kif/Lamborghini
KL/Lagerfeld
Krazy Krizia/Krizia
LA/Max Factor
Le Jardin d'Amour/Max Factor
L'Orientale/Diparco-L'Oréal
Lou-Lou/Cacharel
Magie Noire/Lancôme
Mémoire/Shiseido
Mon Heure/Mibelle
Moschus Exotic Love/Aok-Nerval
Must/Cartier
Nahema/Guerlain
Narcisse Noir/Caron
Nirmala/Molinard
Nuit de Noël/Caron
Nuit D'Eté/Joop
Nuits Indiennes/Jean-Louis
Scherrer
Number Six/Betrix
Opium/Yves St Laurent
Pagan/Jovan
Poison/Christian Dior
Poivre/Atkinsons
Prélude/Balenciaga
Réplique/Raphael
Roma/Betrix-Revlon
Royal Secret/Monteil
Rubis/de Varens

Samsara/Guerlain
Sarabé/Juvena
Scherrer II/Jean Louis
Scherrer
Shalimar/Guerlain
Shocking/Schiaparelli
Siriam/Mibelle
Strategy/Mary Chess
Sublime/Jean Patou
Tabu/Dana
Tapestry/Mary Chess
Tocade/Rochas
Tuvara/Tuvaché
Ultima II/Revlon
Vol de Nuit/Guerlain
Vu/Lapidus
Woman/Jovan
Woodhue/Fabergé
Xia-Xiang/Revlon
Youth Dew/Estée Lauder
Yram/Mary Chess
Zadig/Pucci

LEATHER AND TOBACCO PERFUMES

Aromatics Elixir/Clinique
Cuir de Russie/Chanel
Scandal/Lanvin
Tabac Blond/Caron

FOUGÈRE PERFUMES

Ambush/Dana
English Fern/Penhaligon's
Flor de Blason/Myrurgia
Fougère Royale/Houbigant
Herbissimo Mediterranean
Marjoram/Dana
Herbissimo Mountain Juniper/Dana
Khasana/Coty
Love's Baby Soft/Love
Zany/Avon

MASCULINE FRAGRANCES

GREEN PERFUMES

Adidas/Astor-Beecham
Ajonc/Rocher
Anthracite for Men/Jacomo
Basile Uomo/Sirpea

Biagiotti Uomo/Biagiotti
Bjorn Borg/Borg
Blenheim Bouquet/Penhaligon's
Bowling Green/Beene
Caractère/Hechter

Chevignon pour Homme/Bogart
Devin/Aramis
Eau de Sport/Rabanne
Eau Fraiche/Marbert
Edition/Dunhill

Fahrenheit/Christian Dior
Fair Play/Cerruti
Gentilhomme/Weil
Gold/Yardley
Golf Masters/Castillac

Golf Masters / Castillac
Grey Flannel / Beene
Halston 1–12 / Halston
Henry M. Betrix Country / Betrix
Herbal for Men Old Spice / Shulton
Hero / Fabergé
Insensé Ultramarine / Givenchy
Jovan Grass Oil / Jovan
Knize Two / Knize
Kouros Eau de Sport / Yves St Laurent
Lamborghini Convertible / Lamborghini
Land / Lacoste
Man IV / Jil Sander
Monsieur Lanvin / Lanvin
Moschino / Euroitalia
Motor Racing / Segura
New West / Aramis
Nino Cerutti / Cerruti
Oltas / Lion
Polo Crest / Ralph Lauren
Prestige Dry Herb / Wolff & Sohn
Red for Men / Giorgio Beverly Hills
Sergio Tacchini / Tacchini-Morris
Spring Water / Coryse Salomé
Sybaris / Puig
Tactics / Shiseido
Tech 21 / Shiseido
Thor / Castillac
Trussardi Uomo / Trussardi
Ulric / de Varens
Varens for Men / de Varens
Varens Rouge / de Varens
Varens Vert / de Varens
6–0 / Bjorn Borg

CITRUS PERFUMES

Agua Profunda / Puig
Aqua Velva / Williams
Armani pour Homme / Armani
Auslese / Shiseido
Blue Stratos / Shulton
Bravas / Shiseido
Canoé Sport / Dana
Capucci pour Homme / Capucci
Dimensione Uomo / Ciccarelli
Eau de Cologne Hermès / Hermès
Eau de Monsieur Balmain / Pierre Balmain

Eau de Jean Patou / Jean Patou
Eau de Sport Santos / Cartier
Eau d'Hadrian / Goutal
Eau de Caporal / L'Artisan
Eau Légère / Bourjois
Eau Sauvage / Christian Dior
Eau Sauvage Extrême / Christian Dior
E de C Impériale / Guerlain
E de C Tradition / Elizabeth Arden
English Lavender / Yardley
English Leather / Mens
Fendi Uomo / Fendi
For Men / Raphael
Glacier / Jovan
Gran Valor / Maurer & Wirtz
Green Water / Fath
Hattric Extra Dry / Olivin
Hidalgo / Myrurgia
Hurlingham / Atkinsons
Iceberg Homme / Eurocos
Impériale / Guerlain
Jean Marie Farina / Roger et Gallet
Lacoste / Lacoste-Sofipar
L'Homme / Versace
Lords / Penhaligon's
Mens Club 52 / Helena Rubinstein
Monsieur de Givenchy / Givenchy
Monsieur de Rauch / de Rauch
Napoléon / Juper
Nobile / Gucci
Ocean Rain / Valentino
Or Masculin / Bourjois
Pierre Cardin / Pierre Cardin
Pitralon / Lingner and Fisher
Prestige for Men Cool Frost / Wolff & Sohn
Pyn's Colonia / Parera
Rafale pour Homme / Molinard
Royale Ambrée / Legrain
Royall Lyme / Royall Lyme
Shendy / Roger et Gallet
Sicilian Lime / Crabtree & Evelyn
Signor / Victor
Signoricci / Nina Ricci
Skin Bracer Dry Lime / Mennen
Sport Fragrance / Aigner
Sporting Club / Coryse Salomé
Tailoring for Men / Clinique-Lauder
Tanamar / Parera

Trophée / Lancôme
West Indian Cologne / Crabtree & Evelyn
Windjammer / Avon

LAVENDER PERFUMES

Acqua di Selva / Victor
Agua Brava / Puig
Agua Lavanda / Puig
Cool Sage / Avon
Eau de Balenciaga / Balenciaga
Eau de Lavanda / 4711
English Lavender / Atkinsons
English Lavender / Yardley
Grès pour Homme / Grès
Halston 101 / Halston
Lavande Royale / Legrain
Lavender / Avon
Monsieur F / Ferragamo
Pino Silvestre / Vidal
Silvestre / Victor
V by Victor / Victor
Yachtman / Mas

SPICY PERFUMES

Antarctic / Yves Rocher
Armateur / Payot
Bill Blass for Men / Bill Blass
Borsalino / Borsalino
Bravas Bosky / Shiseido
Brisk Spice-Wild Spice / Avon
Cacharel pour L'Homme / Cacharel
Cameron / Kensington
Chevalier / d'Orsay
Derrick / Orlane
Eau de Lanvin / Lanvin
Equipage / Hermès
Executive / Atkinsons
Francesco Smalto / Smalto
Ginseng for Men / Jovan
Grès Monsieur / Grès
Habit Rouge / Guerlain
Ho Hang / Balenciaga
J. Casanova pour Homme / Casanova
Jacomo / Jacomo
Jaguar for Men / Frey
Jazz / Yves St Laurent
J.H.L / Aramis
Lagerfeld for Men / Lagerfeld
Lapidus pour Homme / Lapidus

Mondo / Mibelle
Marc Cross / Cross
Men's Cologne / Pierre Cardin
Monsieur Carven / Carven
Monsieur Jovan / Jovan
Monsieur Note Havane / Roger et Gallet
Montana pour l'Homme / Montana
Musk Men Free / Aok-Nerval
Old Spice / Shulton
Partner / Révillon
Piment / Payot
Rockford / Atkinsons
Royall Bay Rhum / Royall Lyme
Royall Spice / Royall Lyme
Rugger / Avon
Sex Appeal / Jovan
Squash / Dana
Super Fragrance for Men / Aigner
Team / Femia
Ténéré / Paco Rabanne
Terre de Sud / M. Klein
Trance / Betrix
Trazarra-Nexus / Avon
Valcan / Kanebo
Versailles pour Hommes / Jean Desprez
Weekend / Avon

FLORAL PERFUMES

Antarctic / Yves Rocher
Burley / Armour
Cool Water / Davidoff
Cougar / Yardley
Dunhill / Dunhill
Eau D'Issey Men / Issey Miyake
Eroica / Kanebo
Escape for Men / Calvin Klein
Gilvan / Kanebo
G-Mans / Gainsborough
Hammam Bouquet / Penhaligon's
Jovan Musk Oil / Jovan-Yardley
KL Homme / Lagerfeld
Lauder for Men / Estée Lauder
M / Morabito
Men's Club / Helena Rubinstein
Onyx / Lenthéric
Pour Monsieur / Chanel
Racquet Club / Mem
Raffles / Fine Fragrances

Royal Eroica / Kanebo
Tabac Original / Maurer & Wirtz
The Baron / Evyan
Weil pour Homme / Weil
Wind Drift / Mem
XS / Paco Rabanne

CHYPRE PERFUMES

Acteur / Azzaro
Adolfo for Men / Denney
Agresste / Gal
Alain Delon / Alain Delon
Andron for Men / Jovan
Aqua Cologne Ice Blue / Williams
Aramant / Biodroga
Aramis / Aramis
Babor for Men / Babor
Balestra pour Homme / Balestra
Black Belt / Coty
Black Suede / Avon
Boss No. 1 / Boss
Braggi / Revlon
Brummel / Williams Hispania
Burberrys / Burberry
Cambridge / Mem
Captain / Molyneux
Chaps / Warner
Chromatics / Aramis
Courrèges Homme / Courrèges
Cuirasse / d'Auvillers
Denim / Elida Gibbs
Derringer / Sans Soucis
Drakkar Noir / Guy Laroche
Eau Cendrée / Jacomo
Eau de Sport / Eckstein
Eccelente Man / Alcina
Etienne Aigner No. 1 / Aigner
Gambler / Jovan
Ghibli / Atkinsons
Gianfranco Ferré for Man / Ferré-de Silva
Gucci pour Homme / Gucci
Halston Z14 / Halston
Hardy Amies / Fine Fragrances
Hawk / Mennen
Henry M Betrix Party / Betrix
Héritage / Guerlain
Hud / Avon
Iquitos / Alain Delon
Jean-Charles / Castelbajac-4711

Kanon / Scannon
L'Homme / Roger et Gallet
Lord de Molyneux / Molyneux
Mahe of Shore / Victor
Men / Mennen
Minotaure / Paloma Picasso
Missoni Uomo / Missoni
Mon Triomphe / Williams
Monsieur Houbigant / Houbigant
Noir / Roberre
Olando / Avon
Oriental Spice / Shulton
Paco / Paco Rabanne
Phileas / Nina Ricci
Polo / Warner
Portos / Balenciaga
Punjab / Capucci
Révillon Pour Homme / Révillon
Rothschild / Rothschild
Royal Copenhagen / Swank
Sagamore / Lancôme-L'Oréal
Signor Vivara / Pucci
Silver / Aigner
Sir Irisch Moos / 4711
Stetson / Coty
Temujin / Kanebo
That Man / Revlon
Trouble / Mennen-Revlon
Tsar / Van Cleef & Arpels
Verlande / Gillette
Vintage / Shiseido
Wall Street / Victor-de Modrone
YSL pour Homme / Yves St Laurent
20–21 / Fabergé
Zizanie / Fragonard

FOUGÈRE PERFUMES

Accomplice / Coty
Alvaros / Mibelle
Azzaro pour Homme / Azzaro
Bally Masculin / Bally
Blue Stratos / Shulton
Bogart / Bogart
Boss Sport / Boss
Brando / Parera
British Sterling / Speidel
Brut / Fabergé
Calvin / Calvin Klein
Canoé / Dana
Captain / Molyneux

Casablanca / Coty
Cellini / Fabergé
Chaz / Revlon
Classic Gold / Yardley
Colorado Sage / Bell
Contour / Gillette
Cool Kevin / Mibelle
Der Mann / Jade
Dollar / Coveri
Elements / Hugo Boss
Fer / Feraud-Avon
Fougère Royale pour Homme / Houbigant
Gucci Noble / Gucci
Hai Karate / Leeming-Paquin
Hall Mark / Lenthéric
Imperial Leather / Cussons
Jade East / Swank
Java / St Pres
John Weitz / Weitz
Jordache Man / Jordache
Jules / Christian Dior
Juvena Men / Juvena
King David / Müller
Kouros / Yves St Laurent
Lamborghini GT / Lamborghini
Loewe para Hombre / Loewe
Lordos / Shiseido
Lorenzaccio / Diparco
Lucky / Mas
Lucky Country / Mas
Macho / Fabergé
Magic Men / Mibelle
Mandate / Shulton
Marbert Man / Marbert
Men's Classic / Cantilene-Payot
Men's Style / Juvena
MG5 / Shiseido
Millionaire / Mennen
Ming / Coryse Salomé
Monsieur Rochas / Rochas
Monsieur Worth / Worth
Moustache / Rochas
Old Brown / Parera
Paco Rabanne pour Homme / Paco Rabanne
Patrichs / Philipe
Pitralon Sport / Jovan
Pullmann / Dana
Sir Champaca / 4711

Skin Bracer / Mennen
Skin Bracer Wild Moss / Mennen
Smalto / Smalto
Sport Scent for Men / Jovan
Tai Winds / Avon
Tuscany / Aramis
Un Homme / Jourdan
Version Originale / Sinan
West / Fabergé
Wild Country / Avon
Wings for Men / Giorgio Beverly Hills
Worth pour Homme / Worth
12 / Couturier

WOODY PERFUMES

Aspen / Coty
Balafre / Lancôme
Balenciaga pour Homme / Balenciaga
Bijan for Men / Bijan
Bijan Fragrance for Men / Bijan
Blend 30 / Dunhill
Bois de Vetiver / J. Bogart
Bois Noir / Chanel
Boss Elements / Hugo Boss
Care / Astor
Cerruti 1881 Homme / Cerruti
Colors de Benetton Man / Benetton
Derby / Guerlain
De Viris / J. Bogart
Eau de Monsieur / Venet
Eau de Rochas pour Homme / Rochas
Eau de Vetyver / Yves Rocher
Eau Fringante / d'Orsay
Etienne Aigner No. 2 / Aigner
Ferré Uomo / Ferré
Furyo / Bogart
Gentleman / Givenchy
Giorgio Beverly Hills for Men / Giorgio Beverly Hills
Givenchy / Givenchy
Globe / Rochas
Henry M Betrix City / Betrix
Horizon / Guy Laroche
Kenzo pour Homme / Kenzo
Kipling / Weil
L'Altro Uomo / R. de Camerino
L'Egoiste / Chanel
Macassar / Rochas
Match Play / Bergerat

Men Two / Jil Sander
Metropolis / Estée Lauder
Nomade / d'Orsay
Off Shore / Victor
Oleg Cassini for Men / Cassini
Pagan Man / Yardley
Pour Homme / Balenciaga
Ricci Club / Nina Ricci
Safari for Men / Ralph Lauren
Samarcande / Yves Rocher
Sandalwood / Elizabeth Arden
Santos / Cartier
Tuscany Forte / Aramis
Vetiver / Guerlain
Vetiver / Roger et Gallet
Vetyver Dry / Carven

Vetiver de Puig / Puig
Woodhue for Men / Fabergé
Woods of Windsor for
Gentlemen / Woods of Windsor

LEATHER PERFUMES

Admirals Yachtman / Mas
Antaeus / Chanel
Basala / Shiseido
Camaro / Nerval
Havana / Aramis
Krizia Uomo / Krizia
Matchabelli / Matchabelli
Patrick / Patrick
Pour Lui / Oscar de la Renta
Lancetti Uomo / Lancetti

Lanvin for Men / Lanvin
Léonard pour Homme / Léonard
Man Pure / Jil Sander
Moschino / Moschino
Or Black / Morabito
Pasha / Cartier
Quorum / Puig
Rodeo / Wolff & Sohn
Rouge / Lubin
Snuff / Schiaparelli
Ted / Lapidus
Toro / Marbert
Turbo / Fabergé
Van Cleef & Arpels / Van Cleef &
Arpels
Yatagan / Caron

MUSK PERFUMES

Denim Musk / Elida Gibbs
Light Musk / Avon
Musk Cologne / Jovan
Musk for Men / Avon
Musk for Men / Fabergé
Musk for Men / Nerval
Musk for Men / Yardley
Musk for Men Old Spice / Shulton
Musk Men Soul / Aok-Nerval
Musk Monsieur / Houbigant
Original Musk / Mennen
Royal Copenhagen Musk / Swank
Royall Muske / Royall Lyme
Wild Musk / Coty

BIBLIOGRAPHY

Ackerman, Diane. *A Natural History of the Senses*, Vintage, 1990

Appell, L. *Cosmetics, Fragrances and Flavors*, Novox, 1982

Aykroyd, Bettina. 'Jasmine – La Fleur', *Soap Perfume and Cosmetic*, November 1994

Bedoukian, Paul. 'The Impact of Chemistry on Perfumes – Jicky by Guerlain', *Perfumer & Flavorist*, Vol. 19, January / February 1994

Billot, Marcel. 'A Short History of the Great Perfumers', *Soap Perfume & Cosmetic Year Book*, 1970

Burbidge, F. W. *The Book of the Scented Garden*, The Bodley Head, 1905

Carle, Jean. 'A Method of Creation in Perfumer', *Soap Perfume & Cosmetic Year Book*, 1968

Dayagi-Mendels, Michal. *Perfumes and Cosmetics in the Ancient World*, The Israel Museum, 1993

Devereux, Charla. *The Aromatherapy Kit*, Headline, 1993

Downer, John. *Supersense*, BBC Books, 1988

El Mahdy, Christine. *Mummies, Myth and Magic*, Thames & Hudson, 1989

Fernie, Dr W. T. *Herbal Simples*, John Wright & Sons Ltd, 1914

Forbes, R. J. *A Short History of the Art of Distillation*, E. J. Brill, Leiden, 1970 (1948)

Groom, Nigel. *The Perfume Handbook*, Chapman & Hall, 1992

Jellinek, J. Stephan. 'ESOMAR Seminar Reviews Fragrance Consumer Research', *Perfumer & Flavorist*, Vol. 19 January / February 1994

Kennett, Frances. *History of Perfume*, George G. Harrap & Co., 1975

Kirk-Smith, Dr M. D. 'Human Olfactory Communication', *Ethnological Roots of Culture*, Ed. Gardner, A. Klewer Academic Publishers, 1994

Kirk-Smith, Dr M. D. and Booth, Dr D. A. 'Chemoreception in Human Behaviour: Experimental Analysis of the Social Effects of Fragrances', *Chemical Senses*, Vol. 12, no. 1, 1987

Le Gallienne, Richard. *The Romance of Perfume*, Richard Hudnut, 1928

McKenzie, Dr Dan. *Aromatics and the Soul*, William Heinemann, 1923

Manniche, Lise. *An Ancient Egyptian Herbal*, British Museum Publications Ltd, 1989

Merrin, Archibald C. 'A History of Perfumery', *Perfumery & Essential Oil Record*, Special issue 1927

Müller, Julia. *The H&R Book of Perfume*, Johnson Publications, 1984

Nadim, A. Shaath and Nehla, R. Azzo, 'Egyptian Jasmine', *Perfumer & Flavorist*, Vol. 17, Sept / Oct 1992

Piesse, G. W. Septimus. *The Art of Perfumery*, Longman, Brown, Green, Longmans & Roberts, 1856

Paughe, John (Ed.). *The Physicians of Myddfai*, A facsimile reprinted by Llanerach, 1993

Poucher, W. A. *Perfumes, Cosmetics and Soaps*, Chapman & Hall, 1959

Rimmel, Eugène. *The Book of Perfumes*, Chapman & Hall, 1865

Rostron, P. R. 'Pomanders', *Perfume & Essential Oil Review*, June 1963

Sagarin, Edward. *The Science and Art of Perfumery,* McGraw Hill Book Co, 1945

Samson, Julia. *Nefertiti and Cleopatra,* The Rubicon Press, 1985

Schweisheimer, Dr W. 'A Rose is a Rose, is a Rose Food, and a Rose Perfume and a Rose Medicine', *P &F O R,* March 1954

Schweisheimer, Dr W. 'Animal Odours and Perfumes', *P & F O R,* February 1956

Sommerville, Barbara, Gee, David and Averill, June. 'On the Scent of Body Odour', *New Scientist,* 10 July 1986

Stoddart, Michael D. *The Scented Ape,* Cambridge University Press, 1991

Thompson, C. J. S. *The Mystery and Lore of Perfume,* The Bodley Head, 1927

Tributsch, Helmut. *When the Snakes Awake,* MIT Press, 1982

Unterman, Alan. *Dictionary of Jewish Lore & Legend,* Thames & Hudson, 1991

Van Toller, Steve and Dodd, George H. *Perfumery,* Chapman & Hall, 1988

Watterson, Barbara. *The Gods of Ancient Egypt,* Batsford, 1984

Wilkinson, Richard H. *Symbol and Magic in Egyptian Art,* Thames & Hudson, 1994

Useful Addresses

Butterbur & Sage Ltd (UK)
7 Tessa Road
Reading RG1 8HH
England

Butterbur & Sage Ltd (USA)
894H Route 52
Beacon, NY 12508
USA

Further supplies of the perfume complexes in this kit, plus a full range of therapy quality, Aromark guaranteed essential oils and carrier oils, can be obtained from these addresses.

British Society of Perfumers
Glebe Farmhouse
Mears Ashby Road
Wilby Wellingborough
Northants NN8 2UQ
England

A working body of British perfumers who meet to discuss events taking place in the perfume industry.

International Fragrance Association
8 Rue Charles Humbert
CH1025 Genéve
Switzerland

This is a regulatory body of the fragrance industry.

Index

ACKNOWLEDGEMENTS

First and foremost we greatly appreciate the assistance of Rhona Wells whose expertise was invaluable, not only for the chapter dealing with perfume classifications, but also for the part she played in initially deciding on the perfume complexes and experimenting with them to create the unique formulas provided in Chapter Five.

Thanks are also due to Michael Kirk-Smith for his valuable research work, and Sandy Sandhu. Special thanks go to my husband Paul, who took time out to read through the manuscript despite his own busy schedule.

We also wish to thank those perfume houses who kindly took the time to send us information at such short notice. They are: France Beauté AB, Cassini Perfumes Ltd, Dana Perfumes Corp, Floris of London, Guerlain–Paris, Kenzo Parfums, Lardenois, Mibelle AG Cosmetics, Parfums International Ltd, Jean Patou Perfumes, Paco Rabanne–Paris, Nina Ricci–Paris, Rochas, Woods of Windsor.

EDDISON·SADD EDITIONS
Project Editor   Zoë Hughes
Editor   Marilyn Inglis
Editorial Assistant   Sophie Bevan
Proofreader   Nikky Twyman
Indexer   Dorothy Frame
Art Director   Elaine Partington
Senior Designer   Pritilata Ramjee
Assistant Designers   Lynne Ross and Rachel Kirkland
Illustrator   Julie Carpenter
Production   Charles James and Hazel Kirkman